PREPARING FOR THE REVOLUTION

Information Technology and the Future of the Research University

Panel on the Impact of Information Technology
on the Future of the Research University

Policy and Global Affairs
NATIONAL RESEARCH COUNCIL
OF THE NATIONAL ACADEMIES

THE NATIONAL ACADEMIES PRESS
Washington, D.C.
www.nap.edu

THE NATIONAL ACADEMIES PRESS • 500 Fifth Street, N.W. • Washington, DC 20001

NOTICE: The project that is the subject of this report was approved by the Governing Board of the National Research Council, whose members are drawn from the councils of the National Academy of Sciences, the National Academy of Engineering, and the Institute of Medicine. The members of the committee responsible for the report were chosen for their special competences and with regard for appropriate balance.

This material is based upon work supported by the Woodrow Wilson National Fellowship Foundation, the National Science Foundation (Grant No. EIA-0102264), the W.K. Kellogg Foundation (Grant No. P0085457), and the National Research Council. Any opinions, findings, conclusions, or recommendations expressed in this publication are those of the authors and do not necessarily reflect the views of the organizations or agencies that provided support for the project.

International Standard Book Number 0-309-0864-X

Additional copies of this report are available from the National Academies Press, 500 Fifth Street, N.W., Lockbox 285, Washington, DC 20055; (800) 624-6242 or (202) 334-3313 (in the Washington metropolitan area); Internet, http://www.nap.edu

THE NATIONAL ACADEMIES
Advisers to the Nation on Science, Engineering, and Medicine

The **National Academy of Sciences** is a private, nonprofit, self-perpetuating society of distinguished scholars engaged in scientific and engineering research, dedicated to the furtherance of science and technology and to their use for the general welfare. Upon the authority of the charter granted to it by the Congress in 1863, the Academy has a mandate that requires it to advise the federal government on scientific and technical matters. Dr. Bruce M. Alberts is president of the National Academy of Sciences.

The **National Academy of Engineering** was established in 1964, under the charter of the National Academy of Sciences, as a parallel organization of outstanding engineers. It is autonomous in its administration and in the selection of its members, sharing with the National Academy of Sciences the responsibility for advising the federal government. The National Academy of Engineering also sponsors engineering programs aimed at meeting national needs, encourages education and research, and recognizes the superior achievements of engineers. Dr. Wm. A. Wulf is president of the National Academy of Engineering.

The **Institute of Medicine** was established in 1970 by the National Academy of Sciences to secure the services of eminent members of appropriate professions in the examination of policy matters pertaining to the health of the public. The Institute acts under the responsibility given to the National Academy of Sciences by its congressional charter to be an adviser to the federal government and, upon its own initiative, to identify issues of medical care, research, and education. Dr. Harvey V. Fineberg is president of the Institute of Medicine.

The **National Research Council** was organized by the National Academy of Sciences in 1916 to associate the broad community of science and technology with the Academy's purposes of furthering knowledge and advising the federal government. Functioning in accordance with general policies determined by the Academy, the Council has become the principal operating agency of both the National Academy of Sciences and the National Academy of Engineering in providing services to the government, the public, and the scientific and engineering communities. The Council is administered jointly by both Academies and the Institute of Medicine. Dr. Bruce M. Alberts and Dr. Wm. A. Wulf are chair and vice chair, respectively, of the National Research Council.

www.national-academies.org

PANEL ON THE IMPACT OF INFORMATION TECHNOLOGY ON THE FUTURE OF THE RESEARCH UNIVERSITY

JAMES J. DUDERSTADT (Chair), President Emeritus and Millennium Project Director, The University of Michigan

DANIEL E. ATKINS, Professor of Information and Computer Science, and Executive Director of the Alliance for Community Technology, The University of Michigan

JOHN SEELY BROWN, Chief Scientist, Xerox Corporation

MARYE ANNE FOX, Chancellor, North Carolina State University

RALPH E. GOMORY, President, Alfred P. Sloan Foundation

NILS HASSELMO, President, Association of American Universities

PAUL M. HORN, Senior Vice President for Research, IBM

SHIRLEY ANN JACKSON, President, Rensselaer Polytechnic Institute

FRANK H. T. RHODES, President Emeritus and Professor, Cornell University

MARSHALL S. SMITH, Professor, School of Education, Stanford University and Program Officer for Education, Hewlett Foundation

LEE SPROULL, Professor, Leonard N. Stern School of Business, New York University

DOUG VAN HOUWELING, President and CEO, University Corporation for Advanced Internet Development/ Internet2

ROBERT WEISBUCH, President, Woodrow Wilson National Fellowship Foundation

WM. A. WULF, President, National Academy of Engineering

JOE B. WYATT, Chancellor Emeritus, Vanderbilt University

Principal Study Staff:

RAYMOND E. FORNES, Visiting Senior Scientist/Study Director, and Professor of Physics, North Carolina State University (on sabbatical during 2000-2001)

THOMAS ARRISON, Director, Forum on Information Technology and Research Universities

DAVID BRUGGEMAN, Research Associate, Forum on
    Information Technology and Research Universities
EDVIN HERNANDEZ, Senior Program Associate,
    Government-University-Industry Research Roundtable
STEVEN J. MARCUS, Science Editor/Writer

Acknowledgment of Reviewers

This report has been reviewed in draft form by individuals chosen for their diverse perspectives and technical expertise, in accordance with procedures approved by the National Research Council's Report Review Committee. The purpose of this independent review is to provide candid and critical comments that will assist the institution in making its published report as sound as possible and to ensure that the report meets institutional standards for objectivity, evidence, and responsiveness to the study charge. The review comments and draft manuscript remain confidential to protect the integrity of the deliberative process. We wish to thank the following individuals for their review of this report:

Ruth Dickstein, University of Arizona
James L. Flanagan, Rutgers University
Robert Gallagher, Massachusetts Institute of Technology
Gary Miller, Pennsylvania State University
Gregory Moses, University of Wisconsin
Judy Ozbolt, Vanderbilt University
Matthew Pittinsky, Blackboard, Inc.
Douglas Seefeldt, University of Virginia
Melanie Sturgeon, State of Arizona

Although the reviewers listed above have provided many constructive comments and suggestions, they were not asked to endorse the conclusions or recommendations nor did they see the final draft of the report before its release. The review of this report was overseen by William G. Howard, Jr., Independent Consultant, and John D. Wiley, University of Wisconsin-Madison. Appointed by the National Research Council, they were responsible for making certain that an independent examination of this report was carried out in accordance with institutional procedures and that all review comments were carefully considered. Responsibility for the final content of this report rests entirely with the authoring committee and the institution.

Contents

Preparing for the Revolution

Executive Summary

Reflecting their broad interest in the health of America's research enterprise, the National Academies launched a study in early 2000 on the implications of information technology for the future of the nation's research university—a social institution of great importance to our economic strength, national security, and quality of life.

The premise of this study was a simple one. Although the rapid evolution of digital technology will present numerous challenges and opportunities to the research university, there is a sense that many of the most significant issues are not well understood by academic administrators, faculty, and those who support or depend on the institution's activities.

The study—organized under the Policy and Global Affairs Division of the National Research Council, and undertaken during the past two years—had two major objectives:

- To identify those information technologies likely to evolve in the near term (a decade or less) that could ultimately have major impact on the research university.
- To examine the possible implications of these technologies for the research university—its activities (teaching, research, service, outreach) and its organization, management, and financing—and the impacts on the broader higher-education enterprise.

In addressing the second point, the panel examined those functions, values, and characteristics of the research university most likely to change as well as those most important to preserve.

In pursuit of these ends, a panel was formed that consisted of leaders drawn from industry, higher education, and foundations with expertise in the areas of information technology, the research university, and public policy. The study process included several meetings and site visits, a major workshop, and communication by conference call and e-mail (see Appendixes A and B).

Drawing on its own information-gathering activities, as well as on the growing literature that deals with higher education and information technology, the panel reached several conclusions that should help guide the future efforts of the research university and its stakeholders:

1. The extraordinary pace of information-technology evolution is likely not only to continue for the next several decades but could well accelerate. It will erode, and in some cases obliterate, higher education's usual constraints of space and time. Institutional barriers will be reshaped and possibly transformed.

2. The impact of information technology on the research university will likely be profound, rapid, and discontinuous—just as it has been and will continue to be for our other social institutions (e.g., corporations and governments) and the economy.

3. Digital technology will not only transform the intellectual activities of the research university but will also change how the university is organized, financed, and governed. The technology could drive a convergence of higher education with IT-intensive sectors such as publishing, telecommunications, and entertainment, creating a global "knowledge and learning" industry.

4. Procrastination and inaction are dangerous courses for colleges and universities during a time of rapid technological change, although institutions will also need to avoid making hasty responses to current trends. Just as in earlier periods of change, the university will have to adapt itself to a radically changing world while protecting its most important values and traditions, such as academic freedom, a rational spirit of inquiry, and liberal learning.

5. Although we are confident that information technology will continue its rapid evolution for the foreseeable future and may ultimately have profound impacts on human behavior and social institutions such as the research university, it is far more difficult to predict these impacts with any precision. Nevertheless, higher education must develop mechanisms to at least

sense the potential changes and to aid in the understanding of where the technology may drive it.

6. It is therefore important that university strategies include: the development of sufficient in-house expertise among faculty and staff to track technological trends and assess various courses of action; the opportunity for experimentation; and the ability to form alliances with other academic institutions as well as with for-profit and governmental organizations.

Although the study's discussions and workshops explored a number of policy issues—such as the changing environments for funding and intellectual-property protection—that affect how the research university can best utilize information technology, the panel concluded that to offer recommendations for specific policy changes would be premature. Digital technology is evolving so rapidly that an overly prescriptive set of conclusions and recommendations would be in danger of becoming irrelevant soon after the report's publication.

Given that the foreseeable future will be marked by great uncertainty, the panel instead recommends that the research university and its stakeholders develop a continuing dialogue, with national and grassroots components, to help research institutions and the broader higher-education enterprise understand the advances in information technology and address their potential impacts. The dialogue should involve monitoring specific technological changes and the resulting scholarly, educational, and social shifts; identifying crucial issues, challenges, and opportunities; stimulating awareness on the campuses; and identifying action items for further study.

Briefly put, the dialogue should grapple with the many aspects of the question: How will the research university define and fulfill its missions in a twenty-first century characterized by ubiquitous and rapidly evolving digital technology?

The ultimate goal is to expand and strengthen the research university's intellectual resources and institutional infrastructure not only to manage the anticipated transformation but to lead it. This will require a commonality of understanding among members of the university community (administrators, faculty, students), between disciplines, and between the university and its key external constituents (governing bodies, state governments, federal agencies, and foundations). Such a dialogue can help the research university not only to survive the coming era of rapid change as a vital American institution but to fulfill its traditional roles of education, research, and service more effectively and in as yet undreamed-of ways.

1 Introduction

Our society is now being reshaped by rapid advances in information technologies—computers, telecommunications networks, and other digital systems—that have vastly increased our capacity to know, achieve, and collaborate (Attali, 1992; Brown, 2000; Deming and Metcalfe, 1997; Kurzweil, 1999). These technologies allow us to transmit information quickly and widely, linking distant places and diverse areas of endeavor in productive new ways, and to create communities that just a decade ago were unimaginable.

Of course, our society has been through other periods of dramatic change before, driven by such innovations as the steam engine, railroad, telephone, and automobile. But never before have we experienced technologies that are evolving so rapidly (increasing in power by a hundredfold every decade), altering the constraints of space and time, and reshaping the way we communicate, learn, and think.

The rapid evolution of digital technologies is creating not only new opportunities for our society but challenges to it as well,[1] and institutions of every stripe are grappling to respond by adapting their strategies and activities. Corporations and governments are reorganizing to enhance productivity, improve quality, and control costs. Entire industries have been restructured to better align themselves with the realities of the digital age. It is no great exaggeration to say that information technology is fundamentally changing the relationship between people and knowledge.

Yet ironically, at the most knowledge-based entities of all—our colleges and universities—the pace of transformation has been relatively modest in key areas. Although research has in many ways been transformed by information technology, and it is increasingly used for student and faculty communications, other higher-education functions have remained more or less unchanged. Teaching, for example, largely continues to follow a classroom-centered, seat-based paradigm.

Nevertheless, some major technology-aided teaching experiments are beginning to emerge, and several factors suggest that digital technologies may eventually drive significant change throughout academia (Newman and Scurry, 2000; Hanna, 2000; Noble, 2001). Because these technologies are expanding by orders of magnitude our ability to create, transfer, and apply information, they will have a profound impact on how universities define and fulfill their missions. In particular, the ability of information technology to facilitate new forms of human interaction may allow the transformation of universities toward a greater focus on learning.[2]

American academia has undergone significant change before, beginning with the establishment of secular education during the 18th century (Rudolph, 1991). Another transformation resulted from the Land-Grant College Act of 1862 (Morrill Act), which created institutions that served agriculture and industries; academia was no longer just for the wealthy but charged with providing educational opportunities to the working class as well. Around 1900, the introduction of graduate education began to expand the role of the university in training students for careers both scholarly and professional. The middle of the twentieth century saw two important changes: the G. I. Bill, which provided educational opportunities for millions of returning veterans; and the research partnership between the federal government and universities, which stimulated the evolution of the research university. Looking back, each of these changes seems natural. But at the time, each involved some reassessment both of the structure and mission of the university (Wulf, 1995).

Already, higher education has experienced significant technology-based change, particularly in research,[3] even though it presently lags other sectors in some respects. And we expect that the new technology will eventually also have a profound impact on one of the university's primary activities—teaching—by freeing the classroom from its physical and temporal bounds and by providing students with access to original source materials (Gilbert, 1995). The situations that students will encounter as citizens and professionals can increasingly be simulated and modeled for teaching and learning, and new learning communities driven by information technology will allow universities to better teach students how to be critical analyzers and consumers of information.

The information society has greatly expanded the need for university-level education; lifelong learning is not only a private good for those who pursue it but also a social good in terms of our nation's ability to maintain a vibrant democracy and support a competitive workforce.

But while information technology has the capacity to enhance and enrich teaching and scholarship, it also appears to pose certain threats to our colleges and universities (Duderstadt, 2000a; Katz, 1999) in their current manifestations. We can now use powerful computers and networks to deliver educational services to anyone—any place, any time. Technology can create an open learning environment in which the student, no longer compelled to travel to a particular location in order to participate in a pedagogical process involving tightly integrated studies based mostly on lectures or seminars by local experts, is evolving into an active and demanding consumer of educational services.[4]

Similarly, faculty's scholarly communities are shifting from physical campuses to virtual ones, globally distributed in cyberspace. And technological innovations are stimulating the growth of powerful markets for educational services and the emergence of new for-profit competitors, which could also help reshape the higher-education enterprise (Goldstein, 2000; Shea, 2001).

Technological change also has the potential for transforming how the research university accomplishes its social mission. In an increasingly global culture linked together by technology, with no single cultural context to provide a "filter," the role of traditional disciplinary canons is changing.

It is clear that the digital age poses many questions for academia. For example, what will it mean to be "educated" in the twenty-first century? How will academic research be organized and financed? As the constraints of time and space are relaxed by information technology, how will the role of the university's physical campus change?

In the near term it seems likely that the campus, a geographically concentrated community of scholars and a center of culture, will continue to play a central role, though the current manifestations of higher education may shift. For example, students may choose to distribute their college experience among residential campuses, commuter colleges, and online (virtual) universities. They may also assume more responsibility for, and control over, their education.[5] The scholarly activities of faculty will more frequently involve technology to access distant resources and enhance interaction with colleagues around the world. The boundaries between the university and broader society may blur, just as its many roles will become ever more complex and intertwined with those of other components of the knowledge and learning enterprise (Brown and Duguid, 1996).

Thus we must take care not simply to extrapolate the past but instead to examine the full range of options for the future, even though their precise impacts on society and its institutions will be difficult to predict. In any case, we must be ready for disruption. Just as these technologies have driven rapid, significant, and frequently discontinuous and unforeseen change in other sectors of our society, so too will they present university decision makers not only with exciting prospects but a decidedly bumpy ride.

CONTEXT FOR THE STUDY

Given their mandate from Congress to advise the federal government on scientific and technological matters, the presidents of the National Academies (National Academy of Sciences, National Academy of Engineering, and Institute of Medicine) acted on the above concerns. They launched a project in early 2000, through the National Research Council (NRC), to better understand the implications of information technology for the research university. This institution is a key element of the national research enterprise, a prime mover of the economy, and a critical source of scientists and engineers. Its wide range of academic functions also makes it an important model for analysis, with broad applicability elsewhere in the university community.

Primary support for the National Academies project was provided by the National Research Council, with additional support from the W.K. Kellogg Foundation, the National Science Foundation, and the Woodrow Wilson Fellowship Foundation.

Box 1-1: What is a Research University?

The Carnegie Foundation, in its 1994 classification system of colleges and universities, defined a research university as follows:

- Offers a full range of baccalaureate programs.
- Is committed to graduate education through the doctorate.
- Gives high priority to research.
- Awards 50 or more doctoral degrees a year.
- Receives at least $15.5 million a year in federal support.

In its updated 2000 classification, redefined solely on the basis of degrees awarded, the Carnegie Foundation listed 261 doctoral/research universities. As of fall 1998, these institutions enrolled over 4.24 million students (about 28% of total enrollment nationwide). These universities were also the recipients of over $10 billion in federal research funding in FY 1998 (about 88% of all federal research funding for higher-education institutions).

Source: Compiled by NRC staff from Carnegie Foundation, 2001; Duderstadt, 1999; Kushner, 2001; *Chronicle of Higher Education*, various issues.

The project was organized under the Policy and Global Affairs Division of the NRC, with staff and program support from the Government-University-Industry Research Roundtable.

The premise of this study was simple. Although the rapid evolution of digital technology will present numerous challenges and opportunities to the research university, there is a sense that many of the most significant issues are not well understood by academic administrators, their faculty, and those who support or depend on the institution's activities.

The study had two objectives:

• To identify those information technologies likely to evolve in the near term (a decade or less) that could ultimately have major impact on the research university.

• To examine the possible implications of these technologies for the research university—its activities (teaching, research, service, outreach) and its organization, management, and financing—and the impacts on the broader higher-education enterprise.

In addressing the second point, the panel examined those functions, values, and characteristics of the research university most likely to change as well as those most important to preserve.

In pursuit of these ends, a panel was formed consisting of leaders from industry, higher education, and foundations with expertise in the areas of information technology, the research university, and public policy. Since first convening in February 2000, the Steering Committee has held a number of meetings— including site visits to major technology-development centers such as Lucent (Bell) Laboratories and IBM Research Laboratories—to identify and discuss trends, issues, and options. The major themes addressed by these activities were:

- The pace of evolution of information technology.
- The ubiquitous character of the Internet.
- The relaxation of the conventional constraints of space, time, and institution.
- The pervasive character of information technology (the potential for near-universal access to information, education, and research).
- The changing ways in which we handle digital data, information, and knowledge.
- The growing importance of intellectual capital relative to physical or financial capital.

In January 2001 a two-day workshop was held at the National Academies—with the invited participation of about 80 leaders from higher education, industry, and government—to explore possible strategies for the research university and its various stakeholders and to provide input on possible follow-up initiatives. The presentations and discussions of the workshop were videotaped and broadcast on the Research Channel, and they are currently being videostreamed from its web site (programs.researchchannel.com) to help stimulate public discussion. Members of the panel also participated in a discussion of the project at the June 2001 meeting of the Government-University-Industry Research Roundtable.

This report, finalized through a series of conference calls and email exchanges during the second half of 2001, discusses what the panel learned during the study process. Chapter 2 describes the likely near-future of information technology;

Chapter 3 discusses the implications of this technology for the research university; and Chapter 4 summarizes the panel's findings and calls for a continued dialogue between the research university and its stakeholders on these issues.

The panel has tried to maintain a clear and focused presentation of the issues. In a number of places, it makes assertions based on its collective judgment, while taking care to alert readers and appropriately qualify those assertions. Where possible, the report references the growing literature on information technology and education in order to complement the panel's opinions. Yet change is occurring so rapidly there is high risk that any specific assertion made by individual experts or a panel such as this one may be proved wrong within a few years. Indeed, a central theme of the report is that the research university must be prepared to cope with constant shifts and continued uncertainty regarding information technology and its implications.

In addition, while this report focuses on the 261 U.S. doctoral/research universities, one of the inevitable consequences of the march of information technology is that these universities will become much more interconnected with the rest of higher education. Therefore much of the discussion deals with the broader academic context, of which the research university is but one component.

However, in seeking to gain a broad view of the issues facing the research university and information technology, the panel was unable (given the available time and resources) to examine several issues in the depth it would have liked. Therefore some important topics, such as the service mission of the university, are discussed but briefly.

Finally, although its original charge was to provide specific conclusions and recommendations on a range of policy issues— including some, such as the altered funding environment for the research university and the changes to intellectual-property protection wrought by the digital revolution that are spurring legislative actions, roiling campuses, and finding their way to court—the panel ultimately decided that specificity at this point would be inappropriate and premature. Digital technology is evolving so rapidly that an overly prescriptive set of conclusions and recommendations would be in danger of becoming irrelevant soon after the report's publication. However, the priorities for action that the panel identified are in areas that institutions and the overall higher-education enterprise can themselves consider and begin to address. And academia might get some assistance in that regard. The digital revolution will undoubtedly create barriers and opportunities that permit new federal and state approaches to provide significant leverage in helping the research university anticipate and manage change.

2 Technology Futures

The role of digital technology in the evolving knowledge society is comparable to that of the railroad during the Industrial Revolution (Attali, 1992). With the aid of information-technology "tracks"—high-speed computer and telecommunications systems—now interconnecting so much of the world, reaching into the marketplace, government, and our homes and lives, we often learn about events virtually as soon as they occur and we are able to process the information in a myriad of increasingly useful ways.

This extensive network is bringing peoples and cultures together and creating new social dynamics in the process. It is leading to the formation of closely bonded, widely dispersed communities of people united by their interest in doing business or in sharing experiences and intellectual pursuits. New forms of knowledge accumulation are developing, as are computer-based learning systems that open the way to innovative modes of instruction and learning (Brown, 2000). And new models of libraries are exploiting vast amounts of digital data in physically dispersed computer systems that can be remotely accessed by users over information networks (National Research Council, 2000a).

A major frontier over the next one or two decades is certain to be the "user interface" for complex information systems. How can it become a more natural environment that transcends limitations of keyboard, mouse, and screen—moving toward an immersive environment in which attributes of human face-to-face exchange can essentially be captured? Ultimately, "virtual environments," in which we respond to lifelike simulations that are replete with artificially created sights, sounds, and other stimuli, may liberate us from physical restrictions; current targets

"We are going to have a huge shift in the way people access information. . . . Billions of people worldwide are suddenly able to afford basically the same access that we in this room typically enjoy."
—Stuart Feldman, Workshop on the Impact of Information Technology on the Future of the Research University, January 22-23, 2001, Washington, D.C. (programs. researchchannel. com)

for application include medicine as well as distance education (Olsen, 2000; Young, 2000b).

Information technology thus presents significant opportunities for those in the higher-education enterprise who seek new and better approaches to teaching and learning, research, and public service (National Education Association, 2000; Carr, 2000b; Mendels, 1999). However, the effective use of knowledge in such forms may well require a rethinking of many current assumptions about education in general and the research university in particular (Hanna, 2000; Wulf, 1995).

This chapter is intended to provide an overview of information-technology advances that the panel expects to see over the next decade. Two caveats should be kept in mind. First, the focus of the chapter is on anticipated hardware advances. Yet equivalent advances will be necessary in software development. We face major challenges in cracking the "complexity barrier" in software and developing software systems that diagnose, repair, and protect themselves (National Research Council, 2002 and 2000b). Today's large, complex, and critical information systems may involve hundreds of thousands of computers, be based on millions of lines of code, and operate almost continuously, making them more difficult to design and maintain, and vulnerable in unexpected ways. For example, a growing number of large-scale projects have either been cancelled without being deployed or have experienced significant problems in service (National Research Council, 2000b).

The second caveat is to not confuse technological feasibility with commercial and social reality. Changes in technology will be enormous over the next 10 years (not to mention the next 20), and the *rate* of change is increasing. But individuals, as well as social institutions like the university, cannot rapidly change their behaviors. The fact that we are approaching a time in which communication and access to a great deal of the world's information will be possible in an instant and at near-zero cost does not necessarily mean that education, or other sorts of "knowledge work," will have changed at the same rate. While information technology does hold out the promise of enabling advances in the educational, research, and service functions of the research university, realizing this promise will require time and considerable institutional adaptation.

One of the university's greatest challenges, in fact, will be managing the great discrepancy between technological and institutional change and exploiting the new technological capabilities as best it can, while recognizing that the ability to retrieve data is not the same as knowledge and technical facility is not the same as wisdom. Such challenges are addressed in Chapter 3.

• •

AN EXTRAORDINARY EVOLUTION

It is difficult to appreciate just how quickly information technology is evolving. Five decades ago ENIAC, one of the earliest computers, stood 10 feet tall, stretched 80 feet wide, included more than 17,000 vacuum tubes, and weighed about 30 tons. Today you can buy a musical greeting card with a silicon chip that is 100 times faster than ENIAC (Huey, 1994). Moreover, the time between such improvements is rapidly shrinking. A $1,000 notebook computer now has more computing horsepower than a $20-million supercomputer of the early 1990s.

This extraordinary pace of information-technology evolution is not only expected to continue for the foreseeable future but could well accelerate. For example, the newest supercomputers are capable of performing over 35 trillion calculations per second (Normile, 2002; Reuters, 2002). And computers yet a thousand times faster are currently under development for applications such as the analysis of protein folding (McDonald, 2001).

For the first several decades of the information age, the evolution of digital technology followed the trajectory predicted by "Moore's Law"—a 1965 observation by Intel founder Gordon Moore that the density of transistors on a chip doubles every 18 months or so, thereby making it twice as powerful as before (or, alternatively viewed, half as costly). Although this "law" was intended to characterize silicon-based microprocessors alone, it turns out that almost every aspect of digital technology has advanced at an exponential rate, with some technologies moving forward even faster (Wulf, 1995). For example, disk areal density—the number of bits per square inch that can be put on a disk—has been doubling every 12 months in recent years and is expected to continue at that rate over the near future.[6]

In recent years, information technology has been most dramatically driven not by the continuing increase in computing

Box 2-1: Prefixes Used in this Report

Mega- = 10^6, or a million
Giga- = 10^9, or a billion
Tera- = 10^{12}, or a trillion
Peta- = 10^{15}, or a quadrillion

power but rather by the extraordinary growth of bandwidth—the rate at which we can transmit digital data (Feldman, 2001). In the mid- to late 1980s, 300 bit-per-second modems were in wide use; now the local-area networks in our offices and homes communicate at 10-100 *mega*bits per second and the backbone systems for linking regional networks together typically run at gigabit-per-second speeds. With the rapid deployment of fiber-optic cables and optical switching, terabit-per-second networks are just around the corner (Kahney, 2000). According to one market forecast of the next five years, fiber-optic cable will be installed throughout the world at an equivalent rate of thousands of miles per hour, despite the severe spending slump afflicting the telecommunications industry at the time this report was being prepared.[7] Meanwhile, researchers are already experimenting with moving data at speeds of petabits per second.

IBM reports success in the lab with communications in the 8- to 10-petabit range, and plans are already being made to move such bandwidths into the marketplace (McGarvey, 1999). Some Internet service providers expect to be employing them in perhaps three to five years for their internal traffic. For global communications, intercontinental bandwidth has recently increased from a relatively sclerotic 45 megabits to 88-100 gigabits, made possible by new fiber-optic cable laid under the major oceans.[8]

From the average user's point of view, the exponential rate dictated by Moore's Law will drive increases of 100 to 1,000 in computing speed, storage capacity, and bandwidth every decade. At that pace, today's $1,000-notebook computer will, by the year 2020, have a computing speed of 1 million gigahertz, a memory of thousands of terabytes, and linkages to networks at data transmission speeds of gigabits per second.

Put another way, that notebook computer will have astounding processing and memory capacities. Except its electronics will be so tiny as to be almost invisible, and it will communicate with billions of other computers and devices through wireless technology and global networks—what Lucent (Bell) Laboratories calls a "global communications skin" (Lucent, 2000).

• •

AN INTERNET-DRIVEN ECONOMY

While hardware advances are occurring on a clear trajectory, predictions about future applications are more problematic— they have almost always proven to be either too optimistic or too pessimistic. Still, we can be sure that the nature of human interaction with the digital world—and with other humans through computer-based networks—is certainly evolving. New screen displays, such as one that places nine megapixels on the equivalent of a two-page spread, provide resolutions noticeably better than paper. It's no longer a question of enduring mediocre "I'll put up with this screen" resolution, but one of superlative "I would really like to have it" quality. Advances are being made in other products as well. Thin, readable, and flexible electronic books, for example, are considered "in-the-bag" technology for broad commercialization over the next few years, as are "computers on a wristwatch" and "knowledge in your pocket." For example, the Apple iPod already has a 20 gigabyte drive the size of a quarter.

All the while, we are moving beyond the simple text interactions of electronic mail and electronic conferencing to graphical user interfaces (e.g., the Mac or Windows) to voice to video, and next-generation interfaces may use retinal displays—in which lasers paint images directly on the retina of the eye to portray 360-degree immersive environments. With the rapid development of sensors and robotic actuators, touch and action at a distance—already a reality in robot-assisted surgery—may soon be generally available as well.

Thus the world of the user could be marked by increasing technological sophistication. With virtual reality, individuals may routinely communicate with one another through simulated environments, or "telepresence," perhaps delegating their own digital representations—"software agents," or tools that collect, organize, relate, and summarize knowledge on behalf of their human masters—to interact in a virtual world with those of their colleagues. As communications technology increases in power by 100 fold (or more) each decade, such digitally mediated human interactions could take place with essentially any degree of fidelity desired.

Predictions like these may seem like fantasy, but consider the record: the penetration of digital technology into our society has proceeded at a remarkable pace. In less than a decade, the Internet has evolved from a relatively obscure research network to a commercial infrastructure now actively utilized by 61 percent of U.S. households and essentially all of our schools and businesses (Gartner Group, 2001). On the global level, the Internet already connects hundreds of millions of people with one another, and estimates are that by the end of the decade this number could grow into the billions—a substantial fraction of the world's population.[9] Such growth is expected to continue despite, or perhaps as a result of, the recent rude awakenings of e-business investors to the realities of the marketplace.

More uncertain than the technological trajectory is the status of the business environment, which will greatly influence when advanced capabilities reach the marketplace. Specific forecasts should be treated with skepticism. For example, forecasts of the 2004 worldwide e-commerce market made in early 2001 ranged from $1.4 trillion to $10 trillion (Butler, 2001). In addition, the overall U.S. and world economies experienced significant slowdowns during the year prior to publication of this report. Still, even revised market forecasts predict continued growth in information-technology-related industries, and this growth is expected to accelerate as business conditions improve. Although the exact pace is difficult to predict, the clear trend is that much of the growth in business-to-business commerce will be Internet-driven.

Access to computers and the Internet, and the ability to use this technology, are thus becoming increasingly important to full participation in our nation's economic, political, and social life. Furthermore, the transition from phone links to broadband—and, eventually, fiber optics—will transform the current drippy faucet of modem connectivity to a deluge of gigabits-per-second into our homes, schools, and places of work.

According to several estimates, by 2004 the number of Internet-enabled devices in the world, including mobile phones, personal digital appliances (e.g., Palm Pilots), and other devices, will approach or exceed one billion, and these devices will be "asymptotically cheap"—costing only tens, not thousands, of dollars—and inexorably getting cheaper yet (Feldman, 2001; In-Stat/MDR, 2002). Put another way, over the next decade we could move from "giga" technology (in terms of computer operations per second, storage capacities, and data-transmission rates) to "tera" and then "peta" technology—petabit networks, petabyte databases, and petaflop (quadrillion instructions per second) computing for those applications that need it. We will denominate the number of computer servers in the billions, digital sensors in the tens of billions, and software agents in the trillions.

In effect, we will evolve from "e-commerce," "e-government," and "e-learning" to just about "e-everything" as digital devices increasingly become the primary interfaces not only with our environment but with other people.

3 Implications for the Research University

"**A**re these the shadows of the things that *will* be, or are they the shadows of the things that *may* be?" Thus did a terrified Ebenezer Scrooge (in Charles Dickens' *A Christmas Carol*) beseech his supernatural guide after a vision of the "future" that included the worst-case scenario of his own graceless demise.

Scrooge of course came to learn, as we all eventually do, that while it is hard to predict the future with any accuracy, one can actively work to help shape it. Projections are merely possibilities, some more plausible than others, but all depend on how an enormous set of variables—many of them not quantifiable—actually play out, with and without our intervention.

In Chapter 2, we described some of the information-technology advances that the panel anticipates over the next decade or so. Here, we note that while certain trends in the evolution of technology are apparent, it is difficult to assess their impact on social institutions such as universities with any accuracy. Still, we must try to project, and allow for significant variation in what we can and cannot directly affect, as best we can.

This chapter provides an overview of the unprecedented technology-driven challenges currently being faced by higher education, and by the research university in particular. These challenges are sufficiently great that even the worst-case scenario—the end of the university, an institution that has existed for a millennium and truly become "an icon of our social fabric"—appears to some to be a distinct possibility. The reasoning behind such an extreme prediction is that although the university has survived earlier periods of technology-driven social change with its basic role and structure more or less intact, the changes being induced by information technology are different because they alter the fundamental relationship

"Can an institution such as the university, which has existed for a millennium and become an icon of our social fabric, disappear in a few decades because of technology? Of course. If you doubt it, check on the state of the family farm."
–Wm. A. Wulf
(Wulf, 1995)

between people and knowledge. Thus the technology could profoundly reshape the activities of all institutions, such as the university, whose central function is the creation, preservation, integration, transmission, or application of knowledge.

The panel believes that while the university as a physical place is not in danger of disappearing any time soon, it is nevertheless critical for the higher-education community to prepare itself for change.[10] And it must begin to do so by reconsidering the academic culture that sometimes allows the demand for consensus to thwart action and in which consultation is often defined as consent.

It is encouraging that some challenges of information technology are already being addressed by the higher-education enterprise. For example, regular sections on information technology and distance education have been features of *The Chronicle of Higher Education* for some time. In addition, the long list of references for this report and the involvement of not-for-profit education providers (see examples in Box 3-1), as well as for-profit entities, indicate that a great deal of activity has occurred and is continuing. Universities are also working together and with industry in the area of technology standards to enable the broader changes advocated in this report.[11]

However, experts within and outside academia observe that there is still a great deal of complacency in the research university, and that more intensive and structured communication at the national and campus levels is necessary.[12] The university could fare better in the future if it develops mechanisms to sense the changes being wrought by information technology, speculates broadly on possible effects, and then responds accordingly—with carefully considered strategies backed by prudent investments—not just to avoid extinction but to actively cultivate opportunity.

Learning and scholarship do require some independence from society. The research university in particular provides a relatively cloistered environment in which people can deeply investigate fundamental problems in the natural sciences, social sciences, and humanities, and can learn the art of analyzing difficult problems. But the rapid and substantial changes in store for the university—not only those related to information technology—require that academics work with the institution's many stakeholders to learn of their evolving needs, expectations, and perceptions of higher education. For example,

Box 3-1: Organizations and activities related to information technology and the research university

EDUCAUSE (www.educause.edu) is a nonprofit association whose mission is to advance higher education by promoting the intelligent use of information technology. Membership is open to academic institutions, corporations serving the higher-education information-technology market, and other related associations and organizations.

The Forum For The Future of Higher Education (emcc.mit.edu/forum), consisting of academic leaders and scholars from across the country who convene annually, facilitates shared inquiry and collaboration on issues—primarily in economics, strategy, and technology and learning—likely to influence the future of higher education. The Forum sponsors research, presents findings, and disseminates information throughout the higher-education community. It is an independent, nonprofit organization affiliated with Yale University.

Vision 2010 (www.si.umich.edu/V2010/home.html) is a project, hosted at the University of Michigan, that is concerned with how higher education might be transformed by information technology.

The Futures Project (www.futuresproject.org/), hosted by Brown University's A. Alfred Taubman Center for Public Policy and American Institutions, aims to stimulate an informed debate on the role of higher education in the new global society. It is particularly interested in the opportunities and dangers of a global market for higher education, and in the development of policies that ensure a skilled use of market forces to enhance opportunities while minimizing the associated risks.

The Knight Higher Education Collaborative (www.irhe.upenn.edu/knight/knight-main.html), sponsored by the John S. and James L. Knight Foundation, is composed of institutions and state systems of higher education that work together on policy issues of broad interest and importance. The Collaborative is "housed" administratively at the University of Pennsylvania's Institute for Research on Higher Education (IRHE) and builds on the work started by the Pew Charitable Trust Higher Education Roundtable. The IRHE, headed by Dr. Robert Zemsky, publishes the widely read *Policy Perspectives* series, and has convened or facilitated over 250 roundtables since 1986.

OpenCourseWare (OCW) (ocw.mit.edu/index.html/) is an MIT project in which the university will make nearly all materials from its courses freely available on the World Wide Web for noncommercial use. Depending on the particular class or style in which the course is taught, this could include materials such as lecture notes, course outlines, reading lists, and assignments. More technologically sophisticated content will be encouraged.

Source: Compiled by NRC staff from organization web sites.

universities may be obliged to place a far greater emphasis on forming alliances that allow individual institutions not to try to be all things to all people but to focus instead on their unique strengths.

Universities will have to function in a highly digital environment along with other organizations as almost every academic function will be affected, and sometimes displaced, by modern technology. The ways that universities manage their resources, relate to clients and providers, and conduct their affairs will have to be consistent not only with the nature of their own enterprise but also with the reality of "e-everything." As competitors appear, and in many cases provide more effective and less costly alternatives, universities will be forced to embrace new techniques themselves or outsource some of their functions.

In any case, the panel believes that universities should strive to become learning organizations by systematically studying the learning process and re-examining their role in the digital age. This would involve encouraging experimentation with new paradigms of education, research, and service by harvesting the best ideas, implementing them on a sufficient scale to assess their impacts, and disseminating their fruitful results.

Such self-examination and self-improvement by the research university in particular should include the following issues, each of which is analyzed further—not as prognostication but in the spirit of "shadows of the things that may be"—in the remaining sections of this chapter:

- The university's fundamental activities of education and research.
- The preservation and communication of scholarly knowledge.
- The university's basic form, function, and financing.
- The effect of a changing university on the higher-education enterprise generally.

EDUCATION IN TUNE WITH THE TIMES

The explosive march of hardware capability outlined in Chapter 2 is not being matched by related uses in higher education. One indicator of this gap is the reality that more space on

> "Most organizations do a considerable amount of research about their own functioning. I'm sure that IBM, to take an arbitrary example, spends a tremendous amount of money thinking about IBM and how IBM might function better. Not the university."
> —Don Norman, Workshop on the Impact of Information Technology on the Future of the Research University, January 22-23, 2001, Washington, D.C. (programs. researchchannel.com)

the typical undergraduate's hard drive is likely to be devoted to MP3 music-audio files than to material related to classes. Still, the Napster phenomenon, which focused on university students' and other young adults' predilection to collect popular-music recordings at little or no cost, showed the potential power of leveraging today's commodity hardware capabilities. One future test of success for academia is whether it is able to tap this latent power for educational ends.

But although it has been slow in coming, we're beginning to see the impact of information technology on teaching, and it seems to be driven not so much by faculty or administrators but by the learners themselves. A good number of today's young people have spent their early lives amid visual electronic media such as video games, and they often approach learning as a "plug-and-play" experience. They expect—indeed, demand—interaction; and they are unaccustomed to learning sequentially (e.g., to reading the manual). Instead, they're inclined simply to jump in.

It is, of course, important to distinguish between learning facts and learning concepts, between answering simple factual questions and making difficult judgement calls. Just because learning and teaching environments utilize information technology does not mean that pedagogy and the substance of what is being taught are any less important. While many of today's "digital generation" of media-savvy students are open to new approaches, they will still need to think critically when engaging the materials they encounter, whether surfing the Web or scanning the library stacks.[13]

Other students are less comfortable with information technology; indeed, some would see a threat in any challenge to the deeply engrained notion that "true learning" must occur in a traditional classroom environment. Incorporating new technology into teaching will thus require accommodation to varied learning styles (Passig and Levin, 1999).

Yet we envision a future, enabled by information technology and driven by learner demand, in which two of the major (and taken-for-granted) ways of organizing undergraduate learning will recede in importance: the 55-minute classroom lecture and the common reading list. That digital future will challenge faculty to design technology-based experiences based primarily on interactive, collaborative learning. Although these new approaches will be quite different from traditional ones, they may be far more effective, particularly when provided through a media-rich environment (Hanna, 2000). At the same time, information technology can actually enhance traditional norms of higher education—for example, through the "living-learning" paradigm made possible by Web communities that grow around on-line courses (Young, 2002b).

Such changes also imply a different student-faculty relationship than has traditionally been the case. Students may be more involved in the creation of learning environments, working shoulder to shoulder with the faculty just as they do when serving as research assistants. In that context, student and professor alike are apt to be experts, though in different domains.

The faculty member of the twenty-first century university could thus become more of a consultant or coach than a teacher, less concerned with transmitting intellectual content directly than with inspiring, motivating, and managing an active learning process.[14] That is, faculty may come to interact with undergraduates in ways that resemble how they interact with their doctoral students today.

Higher education is already heavily wired, with 90 percent of four-year-college students going online at least once a day (Greenfield Online, 2000). But in keeping with the academy's customary taste for incremental change, it was natural that the earliest applications of information technology on campus should involve the enhancement of traditional courses. For example, electronic mail and computer conferencing were used to augment classroom discussions, while the Internet provided access to original source materials. Meanwhile, the first applications of computer-aided-instruction technology attempted to automate the more routine aspects of learning.

In other words, consistent with its early applications of other technologies, higher education tended to use digital networks simply to repurpose the traditional lecture course for online access (Newman and Scurry, 2000). Similarly, multimedia networks were used simply as an Internet extension of correspondence or broadcast courses to enhance distance learning.

The most dramatic impacts on university education are yet to come—when learning experiences are reconceptualized to capture the power of information technology. Although the classroom is unlikely to disappear, at least as a place where students and faculty can regularly come together, the traditional lecture format of a faculty member addressing a group of relatively passive students is threatened by powerful new tools such as the simulation of physical phenomena, gaming technology, telepresence, and teleimmersion (the ability of geographically dispersed sites to collaborate in real time). Sophisticated networks and software environments can be used to break the classroom loose from the constraints of place and time to make learning available any place, any time, and to any one. The outlines of what will be possible can even now be seen in the real-time collaboration and project-management tools that are becoming common in the corporate environment.

The attractiveness of computer-mediated distance learning, or "distributed learning," is obvious to adult learners whose work or family obligations bar their routine presence on conventional campuses. But perhaps a more surprising application of computer-based distance learning is the degree to which many on-campus students are now using it to augment their traditional education.[15] Broadband digital networks and multicasting can be used to enhance the multimedia capacity of hundreds of classrooms across campus and link them with residence halls and libraries. Electronic mail has already altered faculty-student interactions in fundamental ways; professors are now much more accessible to their students, as well as to the wider world, than was the case just a few years ago. The apparent downside for some is a decline in informal interactions during office hours and other face-to-face settings (Connolly, 2001).

Meanwhile, online learning enrollments are reportedly growing at a 33 percent annual rate, and are expected to reach 2.2 million by 2004.[16] Despite the well-publicized failure of several e-learning initiatives, 150 institutions now offer online undergraduate degrees and an even greater number offer graduate degrees; and a recent report anticipates that the U.S. e-learning market will recover from the dot.com bust and grow to a size of nearly $14 billion by 2004 (Booz Allen Hamilton, 2002). Little wonder that there has been explosive activity in the commercial sector to create both the content and technology that support this enterprise. While the report points out that corporate and specialized professional training present the greatest growth prospects, it also notes the opportunities for companies that help universities expand e-learning's role in campus-based instruction.

Developing and deploying high-quality distributed-learning curricula can be difficult and expensive, however. Creating online courses is considerably more complex than simply posting lecture notes or PowerPoint presentations on the Web or videostreaming the "talking heads" of lecturing professors (Young, 2000a). Nevertheless, faculty around the country are gaining expertise in how to turn traditional courses into distance-learning courses (Carnevale, 2000a and 2000b) and how to blend traditional and on-line elements (Young, 2002b), and they are reaping significant rewards.

But there are barriers to such innovation as well. Young faculty members who may have the best skills to develop new information-technology approaches to teaching are concerned that this work will not be taken into account in tenure and promotion decisions, although this policy may be changing at some institutions (Young, 2002a). With competing demands on their time, even tenured faculty members may not have sufficient incentive to devote time to utilizing information technology in creative ways. In addition, the question of whether individual faculty or the institution owns "courseware," and other intellectual-property issues, are being raised. These are discussed further below.

Meanwhile, universities are increasingly outsourcing much of the technology and expertise necessary for distributed learning from commercial providers, such as Blackboard and WebCT, which produce course-management systems. They also distribute content in partnership with several educational publishers.[17] These activities are, of course, still evolving. Questions such as who is responsible for changing material as the field evolves and the extent and type of instructional interaction required will be answered over time. Issues such as the apparently higher attrition rates for Web-based (as opposed to traditional) courses also need to be addressed (Carr, 2000a).

The development of effective "edutainment" by educational institutions in partnership with the entertainment industry, using professional actors and production methods, is another future possibility. Given the buying power of the target audience, 18- to 25-year-olds with high income potential, the development could be underwritten by commercial sponsors. In a scenario that seems extreme today, large parts of the general undergraduate curriculum could join college football and basketball as commercialized edutainment, facing university leadership with challenges similar to those now being encountered with regard to sports (Duderstadt, 2000b).

Actually, we are beginning to see the emergence of a whole new type of institution—the virtual university. These entities exist only in cyberspace, without campus or perhaps even faculty, solely to provide distributed-learning opportunities. Unburdened by most of the usual academic constraints, such virtual universities can experiment with a variety of new forms. Some, such as Michigan Virtual University (www.mivu.org), serve only as brokers, providing marketing channels that allow traditional colleges and universities to be "suppliers" of educational services to a distributed marketplace. Others, such as the University of Phoenix, attempt to provide a more complete array of higher-education offerings, including instruction, library support, and administrative services. Its continuing success in transferring its profitable model of part-time education for working adults into cyberspace constitutes an example to other entities (for-profits and non-profits) seeking to reach the same audience.

There are examples of companies creating online universities by disaggregating the overall production of educational programs and selectively outsourcing each component.[18] They hire research-university faculties (to determine content), cognitive scientists (to develop pedagogy and courseware), and instructors (to guide students and develop assessment tools to monitor learning). Similarly, the commercial functions of marketing and distribution can also be disaggregated and outsourced.

By whatever route, distributed learning based on computer-mediated paradigms allows universities to push their campus boundaries outward to serve diverse types of learners. It also facilitates new forms of pedagogy more responsive to a knowledge-based society—in which learning becomes a pervasive, lifetime need. Thus the traditional paradigm of "just-in-case" degree-based education may be augmented, or replaced, by paradigms of "just in time" and customized "just-for-you"— whereby learners will have increased responsibility to select, design, and control the learning environment.

But even as the number of students, institutions, and commercial organizations participating in distance education grows, it is not clear which business models or structures will ultimately succeed. During the time that the panel was completing this study, several for-profit distance-education subsidiaries launched by universities either went under or showed clear signs of stress (Carlson and Carnevale, 2001; Blumenstyk, 2001;

> "Our traditional way of thinking—that once we have the students on our campus they're a captive audience— from my point of view is dead. We have started pursuing new and ambitious collaborations with other universities."
> —Richard Larson, Workshop on the Impact of Information Technology on the Future of the Research University, January 22-23, 2001, Washington, D.C. (programs.researchchannel.com)

Shea, 2001). One criticism of current initiatives is that they often involve merely putting classroom offerings online without fundamentally rethinking their approaches (Young, 2001b). Clearly, the notion that distance education through the Internet would generate substantial revenues quickly and easily has been dispelled. At the same time, there is growing institutional interest in fostering creation of nonproprietary, open-course content and management tools (Young, 2001a; Carr, 2001).

In that spirit, it is important to note that in evaluating the progress of information-technology utilization in the research university, the role of distance learning by itself should not be overemphasized. As the discussion at the beginning of this section indicates, the integration of computer and communications technologies into traditional academic structures—and the creation of new structures, as MIT's OCW initiative is attempting—may have more long-term impact than the distance-education industry per se.

The flood of new initiatives in this area launched in the 1999–2000 period has given way to a shakeout during 2001–2002, as some universities have been shuttering or scaling back their dedicated online units. At this point, it appears that the initiatives likely to survive are those that had been established in the mainstream of the university with a clear mission to help modernize the institution.

RESEARCH UNBOUNDED

So, too, is information technology changing the nature of research. The earliest applications, often limited by computer capacity, were directed at relatively simple mathematical problems in science and engineering. Today, available processing power is much less of a constraint; problems that used to require the computational capacity of rooms full of supercomputers can now be tackled with laptop machines.

The rapid evolution of this technology is also enabling scientists to address previously unsolvable problems—custom-designing new organic molecules, analyzing the complex dynamics of the global climate, or simulating the birth of the universe, just to cite a few. In fact, the use of information technology to simulate natural phenomena has created a fourth modality of research, on a par with observation, theory, and experimentation. Moreover, there is erosion in the conventional understanding that some types of research are more amenable to information-technology contributions than others; new database and modeling tools, for example, are unexpectedly changing fields that had previously made little use of computing power.

New types of research organizations, such as "collaboratories" (far-flung networks of researchers and laboratories) are appearing that could not have existed without this new technology (National Research Council, 1993 and 2001a). Recognizing that information technology is a crucial enabler of advances across a wide range of scientific and engineering fields, both new and established, the National Science Foundation is developing a Cyberinfrastructure Initiative to better integrate instruments, sensors, supercomputers, and high-speed communications networks (Trimble, 2001). Such efforts may make it possible for collaboratories to attack large-scale science and engineering problems requiring diverse, multidisciplinary talent.

Actually, some of the most powerful applications of information technology have already begun occurring in the humanities, social sciences, and the arts. Scholars now use digital libraries such as JSTOR (www.jstor.org) or ArtSTOR to access, search, and analyze complete collections of scholarly journals or works of art (Mellon Foundation, 2001). Archeologists are developing virtual-reality simulations of remote sites and original materials, such as papyrus manuscripts, that can be accessed by colleagues throughout the world.

Meanwhile, social scientists are using powerful software tools to analyze massive data sets of materials collected through interviews and field studies. And practitioners of the visual and performing arts are applying technologies that merge various media—fine art, music, dance, theatre, architecture—and exploit all the senses (visual, aural, tactile, even olfactory) to create new art forms and experiences.

Other, more subtle changes in scholarship are occurring that can be related to emerging information technology, which inherently leverages and enhances intellectual span. The process of creating new knowledge is shifting from the solitary scholar to teams of scholars, often spread over a number of disciplines. This technology also provides the tools—based on artificial intelligence or virtual reality, for example—to even augment the production of knowledge itself. For example, the inter-

"Can the research university survive the locomotive of the IT revolution? I think a much better way to frame the question is: how can the highly valued mission of scientific, technological, humanistic-productivity, and human-capital growth enabled by the research university best be augmented and turbocharged by the IT revolution?"
—Tim Killeen, Workshop on the Impact of Information Technology on the Future of the Research University, January 22-23, 2001, Washington, D.C. (programs. researchchannel.com)

disciplinary field of automated scientific discovery is receiving more attention as the number and accessibility of large databases—e.g., the human genome—increases (Darden, 1997). And theorem-proving software is now commercially available (www.transpowercorp.com). Less restricted to the analysis of what has been, we may effectively create what has *never* been—drawing rather more on the creative experience of the artist than on the analytical skills of the scientist.

Of all the research-university roles examined in this chapter, research would appear to be the one that institutions are best prepared to adapt to new realities. Indeed, federally funded university research has played a critically important role in creating and nurturing the very technologies discussed here (National Research Council, 1999). But while the research university may face relatively greater information-technology challenges in teaching, outreach, and management than in research, the research-related challenges are not trivial. Maintaining the federal-government/university partnership as a driver in the pursuit of fundamental knowledge and as an engine of U.S. and global innovation will require strong commitment from both partners. New modes of cooperation across agencies, institutions, and departments may be needed to fund and effectively utilize the cyber-infrastructure that will enable tomorrow's breakthroughs.

Engaging industry as a partner in research is also an issue. As discussed below, the university research enterprise has become more focused on commercialization and the launch of new ventures than in years past. The new thinking about information technology occurring in the university environment could be an additional magnet for industry interest.

These and other issues have global implications. Given the intensely international nature of today's research, with growing collaboration across distance enabled by information technology, the way that the U.S. research university harnesses new technology in the service of science and engineering is critical not only at home. It is bound to affect scholars and institutions around the world.

PRESERVING AND
COMMUNICATING KNOWLEDGE

The preservation of scholarly knowledge is one of the most rapidly changing functions of the university. The computer, or more precisely the "digital convergence" of various media—from print to graphics to sound to sensory experiences through virtual reality—may ultimately have a greater impact on knowledge than the printing press.

Throughout the centuries, the intellectual focal point of the university has been its library, its collection of written works preserving the knowledge of civilization. Today such knowledge exists in numerous forms—including almost literally in the ether, distributed in digital representations over worldwide networks—and it is not just the prerogative of the privileged few in academe but is accessible to many.

For example, the hypertext link is overshadowing the print bibliographic citation, making original source materials available to all via their own computers. But this is only the tip of the iceberg. The distinction between the book and the library may itself become blurred as the Internet evolves into a seamless mesh for probing the world's "collection." Similarly, because knowledge is not inherently compartmentalized, some disciplinary boundaries may actually devolve. Even without the Internet, Albert Einstein maintained that many of the most critical research challenges lay at the intersections of disciplines. Technology is now increasingly in hand for exploring those intersections.

The library is thus becoming less of a collection house and more of a center for knowledge navigation, a facilitator of information retrieval and dissemination. In addition to utilizing the new "library without walls," scholars and students are increasingly able to access sources directly. As with learning, new electronic media allow the formation of spontaneous communities of unacquainted users, linked together in the many-to-many topology of computer networks. Researchers can now follow the work in their specialization on a day-by-day basis through web sites.

These new realities are giving rise to new challenges as well. The archiving of digital materials is one example. Scholars have found themselves in the odd position of being able to read century-old journal articles—the archived originals—yet unable to read their own manuscripts written with obsolete word-processing software or stored on an obsolete storage medium. The management and preservation of information from short-lived magnetic recording media to ensure future accessibility has not yet been comprehensively addressed. Other issues include potential intellectual distortions resulting from the fact that only relatively recent materials (e.g., journal articles) are available online, and the need for universities to carefully manage heightened demand for access to some paper collections stimulated by electronic access.

Scholarship is still characterized and constrained by the publication of research findings, though this system is fast getting competition as a result of new information technologies (Odlyzko, 2000). The resulting confusion has not yet been resolved: traditional scholarly publication, through established (and extraordinarily costly) journals characterized by peer review, is being challenged by less formal Net-based communication that links scholars essentially instantaneously. The central challenge will be to preserve the benefits of the old system, in which the review process provides cohesion to a given field, while taking advantage of the speed and ease of access promised by new media.

But here too, the technology is evolving. For example, web sites are increasingly serving as portals to integrate material of value to particular scholarly pursuits.[19] Ultimately, the most profound changes will involve software agents (Bradshaw, 1997), though such developments lie some years down the road.

The business environment of academic publishing is also an important factor. Publishers have consolidated and sought to create a business model in which online access to journals will drive profits, but they are faced with several high-profile efforts to expand free online access (Vaidhyanathan, 2001). A coalition of research libraries, universities, and other organizations known as the Scholarly Publishing and Academic Resources Coalition (SPARC) is one group seeking to develop common approaches (www.arl.org/sparc).

Meanwhile, our capacity to reproduce and distribute digital information with perfect accuracy at essentially zero cost has shaken the very foundations of copyright and patent law, and it promises to affect notions of intellectual-property ownership altogether. A Princeton professor's lawsuit to ensure the ability to publish research about unscrambling encrypted digital music illustrates how changing notions of ownership and academic freedom are coming into conflict (Foster, 2001). The Uniform Computer Information Transactions Act (UCITA), a contract-law statute for software developed by the National Conference of Commissioners on Uniform State Laws and adopted by several states, has sparked opposition from a range of groups (Foster, 2000).

Indeed, the legal and economic management of university intellectual property is rapidly becoming one of the most critical and complex issues facing higher education (National Research Council, 2001d). Intellectual-property concerns are often mentioned as a barrier to greater faculty activity in developing new uses of information technology, along with the perceived undervaluing of such efforts in tenure evaluations (Carnevale and Young, 1999).

• •

IMPACT ON THE FORM, FUNCTION, AND FINANCING OF THE UNIVERSITY

Just as new forms of teaching, researching, and preserving knowledge are being stimulated by rapidly evolving information technology, so too will the university's organization, management, governance, and relationships between students, faculty, and staff require serious reevaluation and almost-certain change. For example, the new tools of scholarship and scholarly communication will erode conventional disciplinary boundaries, likely extending the intellectual interests and activities of faculty far beyond traditional academic units such as departments or schools (National Research Council, 2001a). This blurring of disciplinary boundaries does not necessarily contradict the growing need for institutions to build and maintain unique strengths, pointed out earlier in this chapter. These core strengths may be in new fields that combine insights from several traditional disciplines.

Beyond driving a restructuring of the intellectual disciplines, information technology could force a significant disaggregation of many traditional university services, ranging from student housing to health care to teaching itself (Massy, 2001; Newman,

2001; Weiland, 2000). Colleges and universities will increasingly face the question of whether they should continue their full complement of activities or outsource some functions to lower-cost and frequently higher-quality providers.

This will pose a particular challenge to faculty, long accustomed to controlling the design of curriculum and supervising the learning environment. Higher education as a cottage industry, in which individual courses are made to order by individual faculty, may not be able to compete much longer in either cost or quality with commodity educational products (Newman and Couturier, 2001).

Similarly, universities will face a major challenge in retaining instructional "mindshare" among their most accomplished faculty. Higher education adapted long ago to the reality of faculty members negotiating release time and very substantial freedom with regard to research activities. There may be new challenges as instructional content becomes a valuable commodity in a for-profit education marketplace (Thompson, 1999). Some would argue that faculty members should be free to contract with outside organizations in developing instructional learningware; such activity is deemed analogous to scholars authoring textbooks and retaining the royalties. Others maintain that institutions have an ownership interest in such intellectual property. Could policies to restrict such activity be acceptable, or enforceable, in the highly competitive marketplace that exists for leading faculty?

It is possible that we'll ultimately see an "unbundling" of faculty and students from the university, with faculty members acting as freelance consultants, selling their services and knowledge to the highest bidder; and students acting as mobile consumers, able to procure educational services from a highly competitive marketplace (Brown, 1996). Even short of this extreme vision, information technology will likely allow at least some research-university and other higher-education functions to be unbundled—and, where useful, rebundled in new ways.

Movement toward this model would pose a number of challenges to institutions. For example, a student is now considered officially educated when he or she has taken the required credits. But the panel believes that significant learning happens in the "white spaces" between courses and classes—in the heady

"I would allege that the change we're facing is truly discontinuous—in organizations adapted to small, incremental, continuous change. It isn't as if the universities have not changed. But when there's a new technology of the magnitude that we're discussing, discontinuity puts additional stresses on the institutions."
—Marye Anne Fox, Workshop on the Impact of Information Technology on the Future of the Research University, January 22-23, 2001, Washington, D.C. (programs. researchchannel.com)

atmosphere of scholarship and debate that permeates the research university. Transcripts of courses taken thus underestimate a student's education. This problem is apt to be exacerbated as we (correctly) push for more flexibility in how, when, and where we learn.

In contrast to the image of "free agent" professors reaping profits from their learningware, there is an alternative scenario in which incentives for faculty to create new information-technology-based approaches to education are too weak. If the business environment for educational software and contentware is not as favorable as some have anticipated, a gifted young professor might be committing professional suicide by spending large amounts of time creating it. This is a particular risk in the research university, where such activities are not currently an advantage in gaining tenure.

The university faces a particular challenge not only in rewarding the creation of new learning environments but also in ensuring a technology-literate faculty in the first place. Some faculty members have not kept pace with technology's evolution, and they are unprepared for the new plug-and-play generation of students. According to a recent survey of senior information-technology administrators by the Campus Computing Project, 40 percent cited "incorporating technology into the classroom" as the most important issue they face; yet only 14 percent said that technology had improved instruction to date (Carlson, 2000).

In earlier times, we would simply wait for a generation of professors to retire before an academic unit could evolve. But in today's fast-paced world, when the doubling time for technology evolution has collapsed to a few years or less, we must look for effective ways to reskill the faculty members whose careers are far from over.

Actually, almost *all* of a university's adults—faculty, staff, administrators, whomever—need to be reskilled in appreciating how today's student is so effortlessly digital across all boundaries (which are rapidly fading). This issue seldom gets serious attention, even though the ubiquitous presence of computers and other electronic devices—hand-held digital assistants and portable telephones, for example—affects student life at least as much as it does academic programs. In fact, students often make little distinction between the two; they see technology as a fundamental aspect of their lives, seamlessly affecting all of its parts, and they take it for granted just as they do the air they breathe. Woe to the university that doesn't grasp this.

But understanding the need is one thing, and paying for it is another. Thus another major challenge to the university is financial. The bill for information technology is growing faster than those of other categories (Olsen, 2001b). For a very large campus, it can amount to hundreds of millions of dollars per year.[20] It is paradoxical that institutions are spending more and more as a given level of hardware capability becomes less and less costly. This trend underlines not only the special challenges faced in higher education but the difficulty that virtually all organizations have experienced in utilizing information technology to improve productivity (Massy and Zemsky, 1995).

Historically, universities have seen technology as a capital expenditure to serve only a select few, and more or less as an experimental tool. It is often paid for with year-end savings and other "budget dust" (Olsen, 2001a). Though times have changed, most universities still do not have a modern and sustainable financial model for investing in information technology; their planning is largely limited to long-term faculty appointments and even longer-term physical facilities. Trying to satisfy constituents' needs for information-technology infrastructure requires very rapid turnover in large-scale investments, and thus an agility not usually found in a budgeting culture.

Nevertheless, some universities are beginning to realize significant cost savings in administrative areas such as purchasing through the effective use of information technology (Olsen, 2002). This shows that it is possible to invest wisely in new capability that delivers concrete benefits. Not surprisingly, the key to meeting this challenge appears to be the creation and careful management of an organizational structure for tapping the new technology.

IMPACT ON THE HIGHER-EDUCATION ENTERPRISE

Coupled with new societal needs—ubiquitous adult education, for example—and economic realities such as erosion of public support (Hebel, 2001; Healy, 1999; Hebel, Schmidt and Selingo, 2001), information technology is likely not only to transform individual institutions, whether research university or non-research university, but to drive a massive restructuring of the whole higher-education enterprise (Duderstadt, 1999). Judging from the makeovers in other sectors of the economy, such as health care, transportation, communications, and energy, we should expect to see mergers, acquisitions, new competitors, and new products in higher education as well. More generally, we may well see the rise of a global "knowledge and learning" industry, in which the activities of traditional academic institutions converge with those of other knowledge-intensive organizations such as telecommunications, entertainment, and information-services companies.

Such convergence is being driven by the increasing importance of human capital to our knowledge-based economy, which depends so heavily on brainpower, ideas, and entrepreneurship (National Research Council, 2001c). Although the employment and economic situation is weak as this report goes to press, it is clear to many business leaders that obtaining, training, and retaining skilled workers are still critical long-term priorities (ITAA, 2001).

The panel agrees with the general assertion that the emergence of "knowledge work" and "knowledge workers" is crucial to the future development of the global economy and society (Drucker, 1999 and 2001). This notion of "knowledge work" encompasses more than activity directly related to information technology per se; it implies a rise—in nearly all sectors of today's workforce—of professionals who depend on and manipulate information almost exclusively.

A key factor at present in pushing higher education toward restructuring is the emergence of aggressive for-profit education providers intent on satisfying this information demand (Goldstein, 2000). Most of these new entrants, such as the University of Phoenix[21] and Jones International University,[22] are now focusing on the adult-education market as well as corporate training (Hanna, 2000). But they also have more expansive goals in mind.

Having invested heavily in sophisticated instructional content, pedagogy, and assessment tools, these providers are well positioned to offer broader educational programs, both at the undergraduate level and in professional areas such as engineering and law. Thus the initial focus of new for-profit entrants on basic adult education is misleading; in the foreseeable future, their capacity to compete with traditional colleges and universities in some areas could be formidable indeed.

To be sure, some forecasts of demand for distance learning in areas such as business education have proven overly optimistic, at least for the near term (Mangan, 2001; Shea, 2001). But clearly the university will lose its monopoly on students, faculty, and resources, and in the absence of bold commercial alliances it is likely to lose market share to for-profit competitors in its traditional areas of strength.

The research university will face particular challenges in this regard. Although rarely acknowledged, the research university relies heavily on cross-subsidies from low-cost, high-profit instruction in general education (e.g., large lecture courses) and low-cost professional training (such as in business administration and law) to support graduate training and research in the science and engineering fields (Newman, 2000; Newman and Couturier, 2001). These high-profit programs are, not coincidentally, very attractive targets for technology-based, for-profit competitors. Their success in the higher-education marketplace could therefore undermine the current business model of the research university and imperil its core activities. This could be a politically explosive issue for some of the state universities as they try to maintain and increase public support from state legislatures.

Further, as a knowledge-driven economy becomes ever more dependent on new ideas and innovation, there will be growing pressures to commercialize the university's intellectual assets—its faculty and students, its capacity for basic and applied research, and the knowledge generated through its scholarship and instruction—which become ever more valuable (Olcott and Schmidt, 2000). Public policy, through federal actions such as the Bayh-Dole Act of 1980, has encouraged the transfer of knowledge from campus to marketplace. But because knowledge can be transferred not only through formal mechanisms such as patents and licensing but also through the migration of faculty and students, there is a risk that the rich intellectual assets of the university will be depleted as support for graduate education and research erodes.

Even with faculty and students remaining in academia, the research university faces particular conflicts in the commercialization arena. While transforming knowledge into public benefit has long been a major component of its mission, expectations for university contributions to regional and national economic development are growing. Universities are thus forming an array of ambitious partnerships with industry, and are doing more to support faculty entrepreneurship (GUIRR, 2000 and 2001).

Yet some decry the growth of commercial forces and incentives on campus (Press and Washburn, 2000) as a threat to the basic values of the university. Moreover, society's experience so far with market-driven, media-based enterprises has not been altogether positive. The experience of the broadcasting and publishing industries suggests that a narrow focus on short-term financial results can lead to mediocrity.

One can imagine a scenario, for example, in which the campus does not disappear but, because of the escalating costs of residential education, becomes priced beyond the range of all but the most affluent. Much of the population would then be limited to lower-cost education via nonresidential learning centers or computer-mediated distance learning. Indeed, critics see the expansion of distance education as the leading edge of a movement to commercialize higher education and "deprofessionalize" the faculty (Noble, 2001).

While the commercial model of the newer for-profit institutions may be a very effective way to meet the workplace-skill needs of many adults, the committee believes that it is not—or at least, not yet—a paradigm suitable to many of the other purposes of the university, including the educational value of direct interaction with excellent teachers. Also, the traditional brick-and-mortar campus has provided a desirable social environment that contributes substantially to student maturation and to growth into participative citizenship.

Thus even though we must be mindful of market forces and willing to respond to them as creatively and substantially as possible, the panel believes that they should not be allowed to dominate and reshape the higher-education enterprise all by themselves. Otherwise, we could well find ourselves facing a "brave new world" in which some of the most important values and traditions of the university have fallen by the wayside.

As we assess these emerging market-driven learning institutions, we must bear in mind the importance of preserving the ability of the university to serve a broader public purpose. While universities teach skills and convey knowledge, they also preserve our cultural heritage and convey it from one generation to the next, perform the research necessary to generate new knowledge, serve as constructive social critics, and provide society with a broad array of knowledge-based services such as technology transfer and health care.

So what should a university of the twenty-first century—one that serves the needs of a knowledge-driven society—be like? In particular, what will be the research university's role in the changing higher-education infrastructure? It would be impractical and foolhardy to suggest precise models; the great and ever-increasing diversity of the U.S. citizenry and workforce makes it clear that there will be many forms of education and many types of institutions serving our country. But a number of themes will almost certainly factor into the higher-education enterprise.

In a series of reports prepared during the latter half of the 1990s, the Kellogg Commission on the Future of State and Land-Grant Universities charted a future course for an important subset of America's research universities (Kellogg Commission, 2001). The Commission examined the range of internal changes and external forces shaping the future of state and land-grant universities, and did not focus particularly on information technology. In recasting the traditional land-grant missions—education, research, and extension—as learning, discovery, and engagement, it urged universities to re-engage with society and play a more extensive role than in the past.

Information technology can clearly help universities realize the Kellogg Commission's vision. One important area, which this present study was not able to examine comprehensively, is emphasis on the extension (or "engagement") mission. Examples of how this may occur can be seen in fields such as social work, where universities are experimenting with new ways to assist practitioners in the field through information technology (Ouelette, 2001).

The panel believes that just as other social institutions have done, universities must become more focused on those they serve. They must transform themselves from faculty-centered to learner-centered entities, becoming more responsive to what students need to learn—whenever, wherever, and however they wish to learn it—rather than simply cater to what faculties wish to teach. This will become a bigger challenge than ever before as information technology greatly increases the size and enhances the diversity of universities' student bodies, and as more students gain access to computers and reliable networks. In this environment, the Internet has the potential to be a "democratizing" force, extending educational opportunities to those currently underserved by traditional colleges and universities.

To meet the needs of a knowledge-driven society, it is clear that very broad access to education is a high priority. One of the important challenges of the research-university community will be to sustain, intellectually and financially, a "culture of excellence" with the required selectivity of faculty and students while at the same time providing the foundation for universal education.

The research university will undoubtedly play a role in meeting the growing demand for cost-effective educational opportunities. This may involve increased cooperation with other components of the higher-education system such as state universities and community colleges, which have long been accomplished providers of affordable education.

In an age of knowledge, lifelong learning is especially critical. The concept of student and alumnus will merge. Our highly partitioned schooling system may well blend increasingly into a seamless web, in which primary and secondary education; undergraduate, graduate, and professional education; on-the-job training and continuing education; and lifelong enrichment become a continuum. In this vision of the future, people will be continually surrounded by and absorbed in learning experiences.

Information technologies are now providing not only the means to create growth-inducing environments throughout the lives of learners; the technologies themselves will be able to learn and grow throughout their own service lives. Increasingly driven by artificial intelligence and genetic algorithms, such systems will be capable of evolving to serve humanity's changing educational needs.

In all, information technology is rapidly becoming a liberating force in our society, not only freeing us from the mental drudgery of routine tasks but also creating new types of learning communities and, more generally, connecting us with one another in ways we never dreamed possible. Higher education must define its relationship with these emerging trends of the digital age in order to adapt, grow, and continue to excel.

4 Choosing the Future: Findings and Options

nformation technology clearly poses many challenges for higher education in general and the research university in particular. But while the challenges are significant, so too are the opportunities to enhance the important social role of these institutions. The panel endeavored to reflect that spirit in this study.

As noted in Chapter 2, we can expect enormous technological changes over the next 10 years, and with an ever-increasing *rate* of change. Yet Chapter 3 observes that individual human beings cannot modify their behaviors with respect to technology as rapidly as the technology itself is changing. Social institutions such as the law and the university have an even greater inertia with respect to exploiting new technology.[23] Academia's greatest challenge, therefore, will be to resolve this great and growing discrepancy. In order to avoid squandering resources, exhausting faculty, and disappointing students, the higher-education community—and particularly the research university— needs to develop agile processes for experimenting with and assessing alternative courses of action.

Some might argue that while other societal institutions have been transformed or made obsolete by information technology, this is no guarantee that the same will happen to the research university. After all, given the important role of this institution in society and the economy, and its political strength at the statehouse and national levels, it has proven able to resist or deflect outside pressures in the past. For example, some might argue that inexhaustible demand for the research university's degree-granting, or "credentialing," role will ensure a steady stream of students regardless of the intrinsic educational quality it offers.

Although the panel is not in a position to prove its arguments beyond all doubt, we note that a significant number of experts, both panel members and others of diverse outlooks, predict that higher education will undergo significant change as a result of information-technology advances (Collis, 2000; Duderstadt, 2000a; Gilbert, 1995; Katz, 1999; Newman and Scurry, 2000; Noble, 2001).

In addition, Chapter 3 discussed several trends affecting higher education that may not trace their origins to digital technology but that the panel expects will be catalyzed by technological change to further transform institutions and educational processes. For example, there is a growing demand for universal, lower-cost, lifelong education tailored to the needs of learners, as contrasted with the more exclusive, expensive, traditionally structured approaches that the research university and other elite institutions have been accustomed to determining themselves and uniquely providing. The panel expects this growing demand will be met in part, perhaps in large part, through the expansion of current for-profit educational providers or the success of new entrants. In any case, technology is helping to enable this shift.

Another trend that has spurred some controversy is the growing linkage between the research university and the commercial world. Concerns raised about this linkage generally focus on the potential compromising of academic research activities (Press and Washburn, 2000). But some critics worry that the recent rush by universities to establish for-profit subsidiaries specializing in distance education is a mechanism for "deprofessionalizing" the faculty (Noble, 2001). This movement, they suggest, could presage a future in which education is "delivered" by information technology and courses are created by teams of adjunct faculty, contract lecturers, and technical helpers rather than by tenured professors, with the courses owned by the university (Noble, 2001).

Whether or not one shares these particular concerns, it is clear that a range of futures is possible for the research university. The panel believes that institutions, working with their constituents, can develop and fulfill a vision for the future in which information technology is a vehicle for sustaining and expanding their core values and missions. The research university can be more effective in education, teach in entirely new ways, reach a wider segment of the U.S. population, and meet the lifelong learning needs of its students. It can create new knowledge at an accelerating rate through new forms of collaboration across institutional and disciplinary lines, while maintaining diversity of thought and academic freedom. It can be more effective at traditional service functions, and make entirely new contributions to the broader society. It can become even more central to the intellectual and social life of our communities and the nation as a whole than it has been before.

Given the pace of technological change, and the non-technological pressures mentioned above, the next decade will be a critical time for individual institutions and for the higher-education enterprise as a whole. While the transformation of the research university is more or less inevitable, it is important that the changes be proactive and the result of serious self-examination. The aforementioned rush to establish for-profit distance-learning subsidiaries, some of which have already expired while others are in trouble, is an example of change hastily undertaken by institutions in reaction to trends of the moment and the fear of "being left behind." We can learn a great deal by examining this experience.

SIGNIFICANT FINDINGS

Given this context, the panel's main findings are as follows:

• The extraordinary pace of information-technology evolution is likely not only to continue for the next several decades but could well accelerate. It will erode, and in some cases obliterate, higher education's usual constraints of space and time. Institutional boundaries will be reshaped and possibly transformed.

• The impact of information technology on the research university will likely be profound, rapid, and discontinuous—just as it has been and will continue to be for our other social institutions and the economy. There are likely to be major technological surprises, comparable in significance to the personal computer in the late 1970s and the Internet browser in 1994, but at more frequent intervals. The future is becoming less predictable.

• Digital technology will not only transform the intellectual activities of the research university (teaching, research, outreach) but will also change how the university is organized, financed, and governed. The technology could drive a convergence of higher education with IT-intensive sectors such as publishing, telecommunications, and entertainment, creating a global "knowledge and learning" industry.

- Procrastination and inaction are dangerous courses for the university during a time of rapid technological change, although institutions will also need to avoid making hasty responses to current trends. Just as in earlier periods of change, the university will have to adapt itself to a radically changing world while protecting its most important values and traditions, such as academic freedom, a rational spirit of inquiry, and liberal learning.

- For at least the near term, meaning a decade or less, the research university will continue to exist in much its present form. But it must devote itself during this interval to anticipating the needed changes, developing appropriate strategies, and making adequate investments if it is to prosper thereafter.

- Over the longer term, the basic character and structure of the research university may be challenged by the technology-driven forces of aggregation (new alliances, for example, and the conversion of the academic marketplace into a global industry) and disaggregation (such as restructuring of the academic disciplines, detachment of faculty and students from particular universities, and decoupling of research and education).

- Although we are confident that information technology will continue its rapid growth for the foreseeable future and may ultimately have profound impacts on human behavior and social institutions such as the research university, it is far more difficult to predict these impacts with any precision. Nevertheless, higher education must develop mechanisms to at least *sense* the potential changes and to aid in the understanding of where the technology may drive it.

- It is therefore important that university strategies include: the development of sufficient in-house expertise among faculty and staff to track technological trends and assess various courses of action; the opportunity for experimentation; and the ability to form alliances with other academic institutions as well as with for-profit and governmental organizations.

DISCOVERING OPTIONS: THE NEED FOR CONTINUED DIALOGUE

Although part of its charge was to make policy recommendations, the panel ultimately decided not to do so in this first phase of activity. One factor in this decision was that information technology is evolving so rapidly that any prescriptive set of conclusions and recommendations could quickly become outdated. Also, the panel was unable to examine the numerous issues bearing on the topic (such as the state and federal funding environment for higher education, intellectual property laws and practices, regulatory and certification issues, information privacy, and information security) with the depth needed for recommending policy changes. The focus of our examination of needs and priorities for action was on what institutions themselves and their broader constituencies (which definitely includes state and federal governments) need to monitor, explore, learn, and understand at this time.

The panel believes that the higher-education community should create ongoing mechanisms for:

- Monitoring technological changes and the consequent scholarly, educational, and social shifts.
- Identifying crucial issues, challenges, and opportunities for the research university and the broader higher-education enterprise.
- Stimulating awareness on the campuses.
- Making recommendations for actions or further studies.

The National Academies has been awarded funding to launch such an effort through the Government-University-Industry Research Roundtable. This process will address the need for monitoring and expanded dialogue not only on campus but at the national level. It will involve technology specialists as well as experts in higher education and state and federal policy makers. Of course, it will also involve faculty from the humanities, social sciences, natural sciences and engineering, and students themselves.

Box 4-1: Examples of key questions and issues that might be addressed through continued national and campus-based dialogue

For Institutions

1. How will e-learning environments affect the need for traditional teacher-centered instruction? How will the residential campus experience be affected? What are the implications for graduate training in the research university, where graduate-student assistants carry a large share of the teaching load?

2. How will information-technology advances affect the ways in which universities tackle major research problems? What new partnerships among institutions and other constituents (e.g., federal agencies, state governments) are needed for efficient development of the necessary tools?

3. How can the research university become more effective in the planning, procurement, and management of IT infrastructure? What operational and management changes are needed? How can the needs of diverse campus constituencies be better anticipated and addressed? What roles should be played by faculty, students, and administrators?

4. What new policies—for example, on intellectual property, copyright, instructional-content ownership, and faculty contracts—does the research university need to reconsider in light of evolving IT?

For the Higher-education Enterprise and its Public Stakeholders

1. How should the research university address the rapidly evolving commercial marketplace for educational services and content—including, in particular, the for-profit and dot.com providers? How should universities grapple with the forces of aggregation—and disaggregation—associated with technology-driven restructuring of the higher-education enterprise? What new alliances are necessary? Will universities be forced to merge into larger units, as the corporate world has done (though not always with great success)? Will they find it necessary to outsource or spin off existing activities?

2. What are the state and national interests in keeping the universities in step with evolving information technology? What changes in state and federal policies, programs, and investments are necessary in order for higher education to flourish in the digital age?

Why is such an an intensive, expanded dialogue necessary? One reason is the complexity and broad range of issues related to information technology and the research university. This panel took a broad look at the technological, institutional, and policy issues. Given the time and resources available, it was able to cover considerable ground but was unable to delve into issues at great depth. Some important topics, such as the impact of information technology on the service and outreach missions of the research university, were treated only superficially. An expanded national dialogue will allow a deeper examination of the broad range of issues.

A second reason for a continuing dialogue is that the research university itself, its internal component groups, and its key external constituents come to the issues with different interests and perspectives. This report has touched on some of those differences. Effective communication and better common understanding will be necessary to effectively manage the consensual change processes in higher education.

Administrators, faculty, and students, for example, come to the issues with different experiences, expectations, and concerns. University governing bodies and state governments are charged with ensuring effective management of the university and responsiveness to the public interest, yet this oversight is necessarily colored by their respective political and institutional interests. Federal agencies, foundations, industry, and the various higher-education associations are key research-university constituents with their own particular perspectives. In this regard, several nonprofit groups and university-based institutes that are focused on the future of higher education or the use of information technology may contribute a great deal to the discussion.

A third reason for a continuing dialogue on information technology and the research university is that the activity can serve as an ongoing mechanism to track technological changes and their implications for universities. Individual institutions would be unlikely to do this systematically on their own.

What would the continued dialogue consist of, and what could it accomplish? One model for the National Academies activity is the Stresses on Research and Education at Colleges and Universities project that was undertaken during the 1990s by the National Science Board and the Government-University-Industry Research Roundtable (NSB-GUIRR, 1994 and 1998;

Texas A&M University, 1994). This effort consisted of several national colloquia and linked campus-based dialogues to explore and address the new stresses on university-based research that appeared in the early 1990s.

Campus dialogues will be an important grass-roots-level component of the new activity. For example, participating institutions will organize structured dialogues to bring together faculty, students, and administrators to discuss challenges and opportunities presented by digital technology and formulate possible responses. Serving as precedent is the Stresses project, in which several of its campus dialogues unexpectedly catalyzed longer-term strategic planning and change exercises at the participating institutions (Texas A&M University, 1994). In this new activity, therefore, building an infrastructure for continued campus dialogue and change will be consciously built into its own exercises.

Periodic national conferences and workshops will be employed for proposing strategies. Standing subgroups might be formed to develop follow-up strategies and actions (including possible alliances). In addition, the Internet will be used to facilitate the dialogues themselves (through the provision of collaborative space on the project's web site) and to encourage regular exchange among the participating institutions and the broader public.

Additional dialogues will be organized among institutional leaders, such as deans, university trustees, and top faculty, and links will be forged with state and national policy makers and industry leaders. The panel believes that the policy dimension is crucial, although it also believes that public discussions and thinking have not advanced to the point where specific policy issues could be addressed in this present report.[24]

Such sustained activity would be aimed at producing specific initiatives and demonstration projects to help research universities develop appropriate strategies for the digital age. Examples include the use of very-high-bandwidth networks (e.g., Internet2) to support new activities such as multicasting and telepresence, novel approaches to using technology to enhance teaching and learning, and innovative approaches to sustainable financing of information-technology infrastructure.

As a result of this three-year project, we expect that the intellectual community studying issues related to information technology and the research university will be enlarged and strengthened; dialogue across institutions, disciplines, and functions will be enhanced; governmental leaders and key foundations will be engaged; and new approaches to change at the campus and national levels will be taking shape.

BREATHTAKING IMPLICATIONS

There is little doubt that the status quo in higher education cannot, and should not, be maintained as this "disruptive" digital technology finds its way into every corner of our society, and in ever more significant ways. Yet while the challenges to the research university will be great, so too will be the potential to enhance the important social role of this institution.

Academics should approach issues and decisions on information technology in that spirit—not as threats but as opportunities. Creative, visionary leaders can respond by guiding their institutions in new directions that reinforce and augment their most critical roles and values. They can use information technology to help their students learn more successfully, their faculty members become better scholars and teachers, and their institutions serve society inclusively and to ever greater effect.

We are on the threshold of a revolution that is making the world's accumulated information and knowledge accessible to individuals everywhere. It has breathtaking implications for us all, but the challenge is particularly great for the academic community. Our mission—our responsibility—is to develop a strategic framework that enables us to understand this extraordinary technology and shape its impact with skill and imagination. If we are successful, the research university can remain a major source of sustenance for a free and spirited democracy, a vibrant intellectual life, a healthy economy, and other national values.

Endnotes

CHAPTER 1

1. One example is the increased ability to disseminate misinformation over the Internet. At the same time, the success of Snopes.com, a web site that debunks urban legends, illustrates how the Internet can aid in establishing the accuracy or inaccuracy of information. In the academic world, information technology tools have been developed to facilitate *and* thwart plagiarism (Foster, 2002).

2. This is not to imply that institutions and their faculty do not currently value learning. Still, at research universities the relative importance put on research and publishing in tenure decisions is significant, and anecdotal evidence indicates that the emphasis on research may be growing (Wilson, 2001). Nevertheless, the potential for change is illustrated by the emergence of courses in which students from several institutions in different parts of the world learn collaboratively (Cogburn, Levinson, Atkins, and Wielbut, 2001).

3. University researchers in a range of fields have been, and continue to be, "lead users" of new technology (Benner, 2001); the Internet, for example, first emerged as a research application of information technology. Similarly, computer networks are used to enhance libraries' intellectual resources, simulate physical phenomena, and link researchers worldwide in virtual laboratories, or "collaboratories"—advanced, distributed infrastructures that use multimedia networks to relax the constraints on distance, time, and even reality (Kiernan, 1999; National Research Council, 1993 and 2001; National Science Board, 2000). In addition, university management and administrative processes have become heavily dependent on information technology.

4. There are many uncertainties about whether and how online students learn differently from face-to-face students (Koch, 1998).

5. Students' greater responsibility and control may be a mixed blessing. For example, under certain conditions online learning may lead to a greater emphasis on the product—the diploma—as opposed to the education process (Lerych, 2001).

CHAPTER 2

6. At the time this report went to press, an illustrative chart could be found on the IBM web site (www.storage.ibm.com/hdd/technolo/grochows/g02.htm).

7. Fiber-optic market consultant KMI Research predicts that carriers will bury over 344 million kilometers (or 214 million miles) of fiber-optic cable over the 2002-2006 period. Dividing this by the number of hours in five years (43,800) yields a rate of nearly 5,000 miles per hour. See KMI Research, 2002.

8. See Teleography, Inc., 2001. The study indicates that trans-Pacific bandwidth capability increased from 14 gigabits to 244 gigabits per second by the end of 2000, with trans-Atlantic capability at about 550 gigabits per second, and U.S.-Latin American bandwidth capability at about 290 gigabits per second. Bear in mind, however, that this is all backbone cable, and not what anyone could dial into.

9. The July 2001 survey by the Internet Software Consortium located over 125 million unique computer "hosts" on the Internet. According to Matrix Net Systems, if the same rate of growth of recent years is sustained, the Internet will cross the 1-billion-host mark in 2005 (Internet Software Consortium, 2001; Matrix Net Systems, 2000).

CHAPTER 3

10. Newman (2000) provides an overview of the challenges that are facing the universities and forcing change, including advances in information technology.

11. For example, the IMS Global Learning Consortium is developing and promoting open specifications for facilitating online distributed-learning activities such as locating and using educational content, tracking learner progress, reporting learner performance, and exchanging student records between administrative systems (www.imsproject.org). The Advanced Distributed Learning initiative is a university-industry-government effort launched by the Department of Defense in 1997 to develop e-learning standardization (www.adlnet.org).

12. The discussion at the January 22-23, 2001 Workshop on the Impacts of Information Technology on the Future of the Research University (www.researchchannel.com) includes perspectives from several experts.

13. Another cohort of learners pushing for change is employed adults.

14. In this area, information technology can help institutions and faculty move in a direction that they are already exploring. (King, 1993; Grasha, 1994)

15. At some institutions, the distinction between adult learners and on-campus students is becoming increasingly blurred as more full-time students blend face-to-face and on-line coursework in order to balance work, family, and academic obligations.

16. See Pethokoukis (2002) who cites proprietary reports from International Data Corp. and Bear Stearns.

17. See www.blackboard.com and www.webCT.com.

18. Unext is an education company that provides online business education and other e-learning products in collaboration with several universities (including Stanford and Columbia). Courses include targeted training programs and professional development, as well as business education. Unext

operates an accredited online university, Cardean University, that offers business-related courses and an MBA degree.

19. An example of such a portal is Stanford's Highwire Press (highwire.stanford.edu).

20. Michael McRobbie (2001), Indiana University's Chief Information Officer, notes that he operates with a $100 million annual budget and is implementing a $200 million five-year strategic plan for IT.

21. The University of Phoenix (www.phoenix.edu) is a private, for-profit entity that provides high-quality education to working adult students. Through innovative avenues such as distance-education technologies, the University is accessible to working adults regardless of their geographical location. It has 107 campuses in the United States and Canada.

22. Jones International University (www.jonesinternational.edu) is a completely online university that offers undergraduate and graduate degrees as well as certificate programs. Started in 1995, it was accredited in 1999 by the Higher Learning Commission, a member of the North Central Association.

CHAPTER 4

23. But the half-life of students' basic technology is diminishing as technological change accelerates. For twenty years or more the entering college freshman bought and used a typewriter. In the late 1980s and early 1990s it was a personal computer. In the mid-1990s e-mail usage grew. The late 1990s saw the advent of the World Wide Web and Napster. Currently, instant messaging is the "hot" technology.

24. Other groups are examining policy issues related to the future of research universities. For example, in 2000 the Kellogg Commission on the Future of State and Land-Grant Universities proposed a "Millenium Partnership Initiative"—a renewed "partnership of federal and state government, colleges and universities, and the private sector to build the technology infrastructure needed to educate and train the twenty-first century workforce."

References

Attali, Jacques. 1992. *Millennium: Winners and Losers in the Coming World Order*. New York: Times Books.

Barua, Anitesh, Pinnell, Jon, Shutter, Jay and Andrew B. Whinston. 1999. "Measuring the Internet Economy: An Exploratory Study." Austin, Texas: Center for Research in Internet Commerce, School of Business Administration, University of Texas at Austin. (cism.bus.utexas.edu/works/articles/internet_economy.pdf)

Benner, Jeffrey. 2001. "A Grid of Supercomputers." *Wired News.* August 9.

Blumenstyk, Goldie. 2001. "Temple U. Shuts Down For-Profit Distance-Education Company." *The Chronicle of Higher Education* 47 (45) July 20.

Booz Allen Hamilton. 2002. *Re-Learning e-Learning.* June. (www.boozallen.com).

Bradshaw, Jeffrey M. ed. 1997. *Software Agents.* Cambridge, Massachusetts: MIT Press.

Brown, John Seely. 2000. "Growing Up Digital," *Change* 32 (2) March: 10-20.

Brown, John Seely and Paul Duguid. 1996. "Universities in the Digital Age," *Change* 28 (4) July: 11-19.

Butler, Steve. 2001. "The eCommerce: B2B Report." *eMarketer.* February.

Carnegie Foundation for the Advancement of Teaching. 2001. *The Carnegie Classification of Institutions of Higher Education*. Menlo Park, California.

Carnevale, Dan. 2000a. "Turning Traditional Courses Into Distance Courses." *The Chronicle of Higher Education* 46 (48) August 4.

Carnevale, Dan. 2000b. "Indiana U. Scholar Says Distance Education Required New Approach to Teaching." *The Chronicle of Higher Education* 46 (27) March 10.

Carnevale, Dan and Jeffrey R. Young. 1999. "Who Owns On-Line Courses? Colleges and Professors Start to Sort it Out." *The Chronicle of Higher Education* 46 (16) December 17.

Carlson, Scott and Dan Carnevale. 2001. "Debating the Demise of NYUonline." *The Chronicle of Higher Education* 48 (16) December 14.

Carlson, Scott. 2000. "Campus Survey Finds That Adding Technology to Teaching Is a Top Issue." *The Chronicle of Higher Education* 47 (9) October 27.

Carr, Sarah. 2001. "MIT Will Place Course Materials Online." *The Chronicle of Higher Education* 47 (32) April 20.

Carr, Sarah. 2000a. "As Distance Education Comes of Age, the Challenge Is Keeping the Students." *The Chronicle of Higher Education* 46 (22) February 11.

Carr, Sarah. 2000b. "Many Professors Are Optimistic on Distance Learning, Survey Finds." *The Chronicle of Higher Education* 46 (43) July 7.

Cogburn, Derrick L., N. Levinson; Daniel E. Atkins, and Vlad Wielbut. 2001. "Human Capacity Building for the Knowledge Economy: Creating Globally Distributed Web-Based Learning Environments for Advanced Graduate Studies in International Affairs." Paper presented at the 2001 Annual Meeting of the International Studies Association, Chicago, Illinois.

Collis, David. 2000. "New Business Models for Higher Education." In *The Internet and the University*. Devlin, Maureen, Larson, Richard and Joel Meyerson, eds. Cambridge, Massachusetts: Educause: 97-116.

Connolly, Frank W. 2001. "My Students Don't Know What They're Missing." *The Chronicle of Higher Education* 48 (17) December 21.

Darden, Lindley. 1997. "Recent Work in Computational Scientific Discovery." Michael Shafto and Pat Langley, eds. *Proceedings of the Nineteenth Annual Conference of the Cognitive Science Society*. Mahwah, New Jersey: Lawrence Erlbaum: 161-166.

Deming, Peter J. and Robert M. Metcalfe. 1997. *Beyond Calculation: The Next Fifty Years of Computing*. New York: Springer-Verlag.

Drucker, Peter. 1999. "Beyond the Information Revolution." *Atlantic Monthly*. 284:4 (October). (www.theatlantic.com/issues/99oct/9910drucker.htm)

Drucker, Peter. 2001. "The Next Society: A Survey of the Near Future." *The Economist* 356 (32) (3 November): 3-20.

Duderstadt, James J. 2000a. *The Future of the University in the Digital Age*. Speech at the University of Wisconsin at Madison. November 29. (milproj.ummu.umich.edu/publications/uw_it_and_university/uw_it_and_university.pdf)

Duderstadt, James J. 2000b. *Intercollegiate Athletics and the American University: A University Professor's Perspective*. Ann Arbor: University of Michigan Press.

Duderstadt, James J. 1999. "New Roles for the 21st Century University." *Issues in Science and Technology* 16 (2) Winter: 37-44.

Feldman, Stuart. 2001. Presentation on "Technology Futures" at the Workshop on the Impact of Information Technology on the Future of the Research University. January 22. (programs.researchchannel.com)

Foster, Andrea. 2002. "Plagiarism-Detection Tool Creates Legal Quandary." *The Chronicle of Higher Education* 48 (36) May 17.

Foster, Andrea. 2001. "Princeton Computer Scientist Sues for the Right to Speak at a Conference." *The Chronicle of Higher Education* 47 (40) June 22.

Foster, Andrea. 2000. "New Software-Licensing Legislation Said to Imperil Academic Freedom. *The Chronicle of Higher Education* 46 (48) August 11.

Gartner Group. 2001. "Gartner Dataquest Survey Shows 61 Percent of U.S. Households Actively Using the Internet." Press release. August 29. (www4.gartner.com/5_about/press_releases/2001/pr20010829b.html)

Grasha, Anthony F. 1994. "A Matter of Style: The Teacher as Expert, Formal Authority, Personal Model, Facilitator, and Delegator." *College Teaching* 42 (4) Fall.

Greenfield Online. 2000. "The Internet is 'Big Man on Campus." (press release). August 7. (www.greenfieldcentral.com).

Gilbert, Steven W. 1995. "Technology and the Changing Academy." *Change* 27 (5) September: 58-62.

Gobbetti, Enrico and Riccardo Scateni. 1998. "Virtual Reality: Past, Present, and Future." In G. Riva, B. K. Wiederhold, and E. Molinari, eds. *Virtual Environments in Clinical Psychology and Neuroscience: Methods and Techniques in Advanced Patient-Therapist Interaction*. Amsterdam, Netherlands: IOS Presss: 3-20.

Goldstein, Michael B. 2000 "To Be [For-Profit] or Not To Be: What is the Question?" *Change* 32 (5) September: 25-31

Gomory, Ralph. 2000. "Internet Learning: Is It Real and What Does it Mean For Universities?" (www.eng.yale.edu/sheff/Gomory_talk.htm)

Government-University-Industry Research Roundtable. 2001a. *Transcript of the Workshop on the Impact of Information Technology on the Future of the Research University*, January 22-23, 2001, Washington, D.C.

Government-University-Industry Research Roundtable. 2001b. "Commercializing University Research: Aligning Incentives and Protecting the Research Enterprise." Summary Brochure. Washington, D.C.: National Academy Press.

Government-University-Industry Research Roundtable. 2000. "Industry-University Research Partnerships: What Are the Limits of Intimacy?" Summary Brochure. Washington, D.C.: National Academy Press.

Government-University-Industry Research Roundtable. 1998. *A Dialogue on Research University Futures, Proceedings of the 1997 Academies Governing Board Symposium.*

Hanna, Donald, ed. 2000. *Higher Education in an Era of Global Competition.* Madison, Wisconsin: Atwood.

Healy, Patrick. 1999. "Growth in State Spending on Colleges Is Likely to Slow Down." *The Chronicle of Higher Education* 45 (48) August 6.

Hebel, Sara. 2001. "Public Colleges Feel Impact of the Economic Downturn." *The Chronicle of Higher Education* 47 (45) July 20.

Hebel, Sara, Schmidt, Peter and Jeffrey Selingo. "As Legislative Sessions Begin, Colleges Gear Up for a Slowdown." *The Chronicle of Higher Education* 47 (17) January 5.

Huey, John. 1994. "Waking Up to the New Economy," *Fortune*. June 27.

Information Technology Association of America. 2001. *When Can You Start? Building Better Information Technology Skills and Careers.* Washington, D.C.

Internet Software Consortium. 2001. *Internet Domain Survey, July 2001.* July. (www.isc.org).

In-Stat/MDR. 2002. *Mobile Internet Access Devices.* June. (www.instat.com).

Joy, Bill. 2000. "Why the Future Doesn't Need Us." *Wired.* April. (www.wired.com/wired/archive/8.04/joy.html)

Kahney, Leander. 2000. "Net Speed Ain't Seen Nothin' Yet." *Wired News.* March 21. (www.wired.com)

Katz, Richard N. ed. 1999. *Dancing with the Devil: Information Technology and the New Competition in Higher Education.* San Francisco: Jossey-Bass Publishers.

Kellogg Commission. 2001. *Returning to Our Roots: Executive Summaries of the Reports of the Kellogg Commission on the Future of State and Land-Grant Universities.* Washington, D.C.: National Association of State Universities and Land-Grant Colleges. (www.nasulgc.org)

Kharaf, Olga. 2001. "The Fiber-Optic 'Glut' in a New Light." *Businessweek Online.*

Kiernan, Vincent. 1999. "Internet-based 'Collaboratories' Help Scientists Work Together." *The Chronicle of Higher Education* 45 (27) March 12.

King, Alison. 1993. "From Sage on the Stage to Guide on the Side." *College Teaching* 41 (1) Winter.

KMI Research. 2002. *Worldwide Cable Market Summary by Region (Single-Mode Plus Multimode).* Private communication. August 15. (www.kmicorp.com).

Koch, James. 1998. "How Women Actually Perform in Distance Education." *The Chronicle of Higher Education* 45 (2) September 11.

Kurzweil, Ray. 1999. *The Age of Spiritual Machines: When Computers Exceed Human Intelligence.* New York: Viking.

Kushner, Mark J. 2001. "Core Values and the New Business Model." *ASEE Prism Online.* May. (www.asee.org/prism/may01/lw.cfm)

Lewis, Laurie, Kyle Snow, Elizabeth Farris and Douglas Levin. 2000. "Distance Education at Postsecondary Institutions, 1997-98." Washington, D.C.: National Center for Education Statistics. nces.ed.gov/pubs2000/qrtlyspring/5post/q5-7.html)

Lerych, Lynne Drury. 2001. "Meeting the Bottom Line in the College Biz." *Newsweek.* April 9.

Lucent Technologies. 2000. "Networking." *Trends and Developments* 4 (2) (www.lucent.com/minds/trends/trends_v4n2/page1.html)

Mangan, Katherine. 2001. "Expectations Evaporate for Online MBA Programs," *The Chronicle of Higher Education* 48 (6) October 5.

Massy, William. 2001. Presentation on "The Impact of IT on the Broader Environment of the Research University" at the Workshop on the Impact of Information Technology on the Future of the Research University. January 22. Washington, D.C. (programs.researchchannel.com).

Massy, William and Robert Zemsky. 1995. "Using Information Technology to Enhance Academic Productivity." Educom White Paper. (www.educause.edu/nlii/keydocs/massy.html)

Matrix NetSystems. 2000. "Internet Survey Reaches 93 Million Internet-Host Level. Press release. September. (www.matrixnetsystems.com).

McDonald, Tim. 2001. "IBM's 'Blue Gene' Supercomputer to Top ASCI White." *Newsfactor Network.* August 22. (www.newsfactor.com/)

McGarvey, Joe. 1999. "Net Experts Predict Petabit Future." *Interactive Week.* September 17.

McRobbie, Michael. 2001. Presentation on "The Impact of IT on Organization and Structure" at the Workshop on the Impact of Information Technology on the Future of the Research University. January 22. Washington, D.C. (programs.researchchannel.com)

Mendels, Pamela. 1999. "Non-Traditional Teachers More Likely to Use the Net." *The New York Times.* May 26.

Mellon Foundation. 2001. Recent Announcements. April 5. (www.mellon.org)

Merrill Lynch. 2000. *The Knowledge Web.*

National Education Association 2000. "Confronting the Future of Distance Learning—Placing Quality in Reach." News release announcing survey results. June 14. (www.nea.org/nr/nr000614.html)

National Research Council. 2002. *Cybersecurity Today and Tomorrow: Pay Now or Pay Later.* Washington, D.C.: National Academy Press.

National Research Council. 2001a. *Issues for Science and Engineering Researchers in the Digital Age.* Washington, D.C.: National Academy Press.

National Research Council. 2001b. *Embedded Everywhere: A Research Agenda for Networked Systems of Embedded Computers.* Washington, D.C.: National Academy Press.

National Research Council. 2001c. *Building a Workforce for the Information Economy.* Washington, D.C.: National Academy Press.

National Research Council. 2001d. *Academic IP: The Effects of University Patenting and Licensing on Commercialization and Research.* Washington, D.C.: National Academy Press.

National Research Council. 2000a. *LC21: A Digital Strategy for the Library of Congress.* Washington, D.C.: National Academy Press.

National Research Council. 2000b. *Making IT Better: Expanding Information Technology Research to Meet Society's Needs.* Washington, D.C.: National Academy Press.

National Research Council. 1999. *Funding a Revolution: Government Support for Computing Research.* Washington, D.C.: National Academy Press.

National Research Council. 1993. *National Collaboratories: Applying Information Technology for Scientific Research.* Washington, D.C.: National Academy Press.

National Science Board and Government-University-Industry Research Roundtable. 1998. *Stresses on Research and Education at Colleges and Universities: Phase II of a Grass-roots Inquiry.* (www7.nationalacademies.org/guirr/stresses.html)

National Science Board and Government-University-Industry Research Roundtable. 1994. *Stresses on Research and Education at Colleges and Universities: Institutional and Sponsoring Agency Responses.*

National Science Board. 1999. Education and Human Resource Committee Workplan, November 18.

National Science Board. 2000. *Science and Engineering Indicators—2000.* Arlington, Virginia: National Science Foundation.

National Science and Technology Council. 2000. *Information Technology: The 21st Century Revolution.* Washington, D.C.: U.S. Government Printing Office.

Newman, Frank. 2001. Presentation on "The Impact of IT on the Broader Environment of the Research University" at the Workshop on the Impact of Information Technology on the Future of the Research University. January 22. Washington, D.C. (programs.researchchannel.com)

Newman, Frank. 2000. "Saving Higher Education's Soul." *Change* 32 (5) September: 16-23. (www.futuresproject.org/publications/soul.pdf)

Newman, Frank and Lara K. Couturier. 2001. "The New Competitive Arena: Market Forces Invade the Academy." *Change* 33 (5) September: 11-17.

Newman, Frank and Jaime Scurry. 2000. "Higher Education in the Digital Rapids," The Futures Project: Policy for Higher Education in a Changing World. (www.futuresproject.org/publications/digitalrapidsreport1.pdf.)

Noble, David. 2001. "The Future of the Faculty in the Digital Diploma Mill." *Academe* 87 (5) September-October.

Normile, Dennis. 2002. "'Earth Simulator' Puts Japan on the Cutting Edge." *Science* 295. March 2.

Odlyzko, Andrew. 2000. "The Future of Scientific Communication." (www.dtc.umn.edu/~odlyzko/doc/future.scientific.comm.pdf)

Olcott, Donald, Jr. and Kathy Schmidt. 2000. "Redefining Faculty Policies and Practices for the Knowledge Age." Donald Hanna, ed. In *Higher Education in an Era of Global Competition.* Madison, Wisconsin: Atwood.

Olsen, Florence. 2002. "E-Commerce Starts to Pay Off." *The Chronicle of Higher Education* 48 (46) July 26.

Olsen, Florence. 2001a. "Report Details Options on Paying for Technology." *The Chronicle of Higher Education* 47 (37) May 25.

Olsen, Florence. 2001b. "Survey Documents Increased Spending by Colleges on Information Technology." *The Chronicle of Higher Education* 48 (11) November 9.

Olsen, Florence. 2000. "Scholars in Medicine and Psychology Explore Uses of Virtual Reality." *The Chronicle of Higher Education* 47 (3) September 22.

Ouellette, Philip and S.P. Sells. 2001. "Creating a telelearning community for training social work practitioners working with troubled youth and their families." *Journal of Technology and Human Services* 20:1-3.

Passig, David and Haya Levin. 1999. "Gender interest differences with multimedia learning interfaces." *Computers in Human Behavior* 12 (2) March.

Pethokoukis, James M. 2002. "E-learn and earn." *U.S. News and World Report.* June 24.

Press, Eyal and Jennifer Washburn. 2000. "The Kept University." *The Atlantic Monthly* 285 (3) March.

Reuters. 2002. "Japan claims fastest computer title." (news.com.com/2102-1001-887718.html)

Rudolph, Frederick. 1991. *The American College and University: A History.* Athens, Georgia: University of Georgia Press.

Shea, Christopher. 2001. "Taking Classes to the Masses," *The Washington Post Magazine* September 18:25-33.

Telegeography, Inc. 2001. *International Bandwidth 2001.* Washington, D.C. (www.telegeography.com/Publications/bandwidth2001/ib2001_exec_summary.pdf)

The Chronicle of Higher Education. 2000. "Definitions of 1994 Carnegie Classifications." 47:13. November 24.

Texas A&M University. 1994. *Final Implementation Report of the Campus-Based Discussions at Texas A&M University.*

Thompson, Dennis F. 1999. "Intellectual Property Meets Information Technology," *Educom Review* 34 (2). (www.educause.edu)

Trimble, Paula Shaki. 2001. "Study Eyes Terascale Road Map." *Federal Computer Week.* March 12.

Vaidhyanathan, Siva. 2001. *Copyrights and copywrongs : the rise of intellectual property and how it threatens creativity.* New York : New York University Press.

Vision 2010 Project web site (www.si.umich.edu/V2010/home.html)

Weiland, Todd. 2000. "Unbundled: Higher Education's Inevitably Virtual Future." *VirtualStudent.com.* July. (www.virtualstudent.com/html/unbundled.html)

Wilson, Robin. 2001. "A Higher Bar for Earning Tenure." *The Chronicle of Higher Education* 47 (16) January 5.

Wulf, Wm. A. 1994. *University Alert: The Information Railroad Is Coming.* (www.cs.virginia.edu/~wulf/documents/buttons.html)

Wulf, Wm. A. 1995. "Warning: Information Technology Will Transform the University," *Issues in Science and Technology* 11 (4) Summer: 46-52.

Young, Jeffrey R. 2002a. "Ever So Slowly, Colleges Begin to Count Work with Technology in Tenure Decisions." *The Chronicle of Higher Education* 48 (23) February 22.

Young, Jeffrey R. 2002b. "The 24-Hour Professor." *The Chronicle of Higher Education* 48 (38) May 31.

Young, Jeffrey R. 2001a. "Universities Begin Creating a Free, 'Open Source' Course-Management System." *The Chronicle of Higher Education* 47 (34) May 4.

Young, Jeffrey R. 2001b. "Logging in with Farhad Saba: Professor Says Distance Education Will Flop Unless Universities Revamp Themselves." *The Chronicle of Higher Education* 47 (42) June 29.

Young, Jeffrey R. 2000a. "Faculty Report at U. of Illinois Casts Skeptical Eye on Distance Education." *The Chronicle of Higher Education* 46 (19) January 14.

Young, Jeffrey R. 2000b. "Virtual Reality on a Desktop Hailed as a New Tool in Distance Education." *The Chronicle of Higher Education* 47 (5) September 22.

Appendixes

A Project Chronology

What follows is a chronology of meetings and activities of the NRC project on the Impact of Information Technology on the Future of the Research University (ITFRU). Additional background can be found on the project web site (www7.nationalacademies.org/guirr).

February 14, 2000
First meeting of Steering Committee ("panel"), Washington, D.C.

May 5, 2000
Meeting of Steering Committee Technology Subgroup on "Cutting Edge IT Issues," Sloan Foundation, New York, N.Y.

June 9, 2000
Steering Committee members testified at the House Sub-committee on Basic Research hearing, "The Internet, Distance Learning, and the Future of the Research University." More information on the hearing, including witness statements, can be found at www.house.gov/science/hearing_106.htm

July 20, 2000
Conference call of the Steering Committee University Subgroup focusing on impacts of information technology on instruction and education

August 17, 2000
Conference call of the Steering Committee University Subgroup

August 24-25, 2000
Meeting of Steering Committee, "A View of Technology Futures," Bell Labs (Murray Hill, N.J.) and IBM Watson Labs (Yorktown Heights, N.Y.)

December 4, 2000
Conference call of the Steering Committee concerning workshop planning

January 16, 2001
Conference call of the Steering Committee to finalize workshop planning

January 22-23, 2001
Workshop on the Impact of Information Technology on the Future of the Research University, Washington, D.C.

March 6, 2001
Broadcasts of the first day's workshop sessions begin on the Research Channel. Sessions are available for viewing at programs.researchchannel.com

April 24, 2001
Conference call of the Steering Committee concerning the January workshop, June GUIRR meeting, and Phase II funding

June 19-20, 2001
Members of the Steering Committee facilitate discussion of the project at the Government-University-Industry Research Roundtable Council meeting

August 29, 2001
Conference call of the Steering Committee regarding the June GUIRR meeting, continuing efforts for Phase II funding, and concluding Phase I activities

Fall-Winter, 2001-2002
Preparation of final project report

B Workshop Agenda January 22-23, 2001

IMPACT OF INFORMATION TECHNOLOGY ON THE FUTURE OF THE RESEARCH UNIVERSITY

CHAIRED BY JAMES J. DUDERSTADT
PRESIDENT EMERITUS: UNIVERSITY OF MICHIGAN

January 22-23, 2001
Washington, DC

January 22, 2001
Lecture Room, The National Academy of Sciences Building

7:45 AM   Continental Breakfast

8:15   Welcome, Introductions, Background and Objectives
         (Jim Duderstadt)

8:30   Wm. A. Wulf,
         President,
         National Academy of Engineering
      Plenary Address:
         The Information Technology Train—
         A Wakeup Call to the Research University

8:45   Technology Futures
      Moderator:
         Dan Atkins
            Executive Director,
            Alliance for Community Technology

Discussants
    Fred Brooks
        Chair, Computer Science Department,
        University of North Carolina, Chapel Hill
    Stu Feldman
        President, IBM Worldwide Computing

10:30   Break

10:45   The Impact of IT on the Activities of the University
            (Teaching, Research, Service)
            Moderator:
                Joe Wyatt
                    Chancellor Emeritus, Vanderbilt University
            Discussants:
                Tim Killeen
                    Director,
                    National Center for Atmospheric Research
                Richard Larson
                    Professor of Electrical Engineering, MIT
                Gary Miller
                    Associate Vice President,
                    Distance Education,
                    Pennsylvania State University
                Don Norman
                    Professor Emeritus,
                    University of California, San Diego

12:30 PM   Lunch

1:30   The Impact of IT on Organization and Structure
            Moderator:
                Nils Hasselmo
                    President,
                    American Association of Universities

Discussants:
> Jon Cole
>> Provost, Columbia University
> Marye Anne Fox
>> Chancellor, North Carolina State University
> Mike McRobbie
>> Vice President for Information Technology/
>> Chief Information Officer,
>> Indiana University
> Barbara O'Keefe
>> Dean, School of Speech,
>> Northwestern University

3:15   Break

3:45   The Impact of IT on the Broader Environment of the Research University (e.g., post-secondary education marketplace, research enterprise)
Moderator:
> Doug Van Houweling
>> President, University Corporation for
>> Advanced Internet Development/Internet2
Discussants:
> Bill Massy
>> President,
>> Jackson Hole Higher Education Group
> Frank Newman
>> Director, The Futures Project
> Diana Oblinger
>> Professor of the Practice,
>> Kenan-Flagler Business School,
>> University of North Carolina, Chapel Hill
> Bob Zemsky
>> Trustee, Franklin and Mills College

5:30   First Day Wrap-up (Jim Duderstadt)

5:45   Reception

January 23, 2001
Members Room, The National Academy of Sciences Building

7:45 AM   Continental Breakfast

8:15   Informal Remarks and Discussion (Wm. A. Wulf)
          Potential Impacts of IT on the Research University
          and Possible Actions

8:45   Breakout groups: "How should the research
          university respond to the challenges, threats, and
          opportunities associated with IT?"

          What should institutions do themselves?
              Moderater: Bob Weisbuch, Members Room
          What should the federal government do?
              Moderater: Dan Atkins, Board Room
          What should industry do?
              Moderater: Lee Sproull, Room 280

10:45   Break

11:15   Breakout Group Reports and Discussion

12:00 PM Lunch

1:00   How best can the National Academies' ITFRU
          Project stimulate and support such actions?
          For example, should we:
          •   Establish an ongoing dialogue that will engage
          campuses?
          •   Organize further workshops or focus groups on
          campus?
          •   Develop a national Web portal on the subject?

3:30   Meeting Wrap-up (Jim Duderstadt)

4:00 PM   Adjourn

C Panel Member Bio Sketches

James J. Duderstadt (*Chair*) is President Emeritus and University Professor of Science and Engineering at the University of Michigan. He also is the Director of the Millennium Project, a research center concerned with the future of higher education. Dr. Duderstadt obtained his B.S. in electrical engineering from Yale and his Ph.D. in engineering science and physics from the California Institute of Technology. He joined the faculty of the University of Michigan in 1968, and served as Dean of the College of Engineering and then Provost and Vice President for Academic Affairs before becoming President of the university in 1988. Dr. Duderstadt's teaching and research interests span a range of subjects in science, mathematics, and engineering, including science policy and higher education. Dr. Duderstadt has received several national awards and has been elected to many honorific societies. He has chaired or served on numerous boards, including the National Science Board, the Executive Council of the National Academy of Engineering, the Committee on Science, Engineering, and Public Policy of the National Academy of Sciences, the Big Ten Athletic Conference, Unisys, and CMS Energy.

Daniel E. Atkins earned a B.S. in electrical engineering from Bucknell University in 1965, and an M.S.E.E. and a Ph.D. in computer science from the University of Illinois, Champaign-Urbana, in 1967 and 1970, respectively. Dr. Atkins joined the University of Michigan's Department of Electrical Engineering and Computer Science (EECS) as an assistant professor in 1972. From January 1989 through July 1990, he served as interim Dean of the College of Engineering. In 1990 Dr. Atkins created an R&D consortium to realize a prototype of a "collaboratory," a vision around which a large and interdisciplinary group of faculty and administrators have coalesced their interests. Dr. Atkins became founding Dean of the new School of Information in July 1992 and held that position until September 1998. With major support of the University and the W. K. Kellogg Foundation, Dr. Atkins led the School of Information's creation of a graduate research and educational program to produce leaders and change agents in the design, use, and evaluation of new knowledge-work environments. Dr. Atkins is currently the Executive Director of the Alliance for Community Technology, a strategic partnership with the W. K. Kellogg Foundation.

John Seely Brown is Chief Scientist of the Xerox Corporation. He has been deeply involved at Xerox in expanding the role of corporate research to include organizational learning, ethnographies of the workplace, complex adaptive systems, and techniques for unfreezing the corporate mind. His research interests include digital culture, ubiquitous computing, user-centering design, and organizational and individual learning. Dr. Brown is a cofounder of the Institute for Research on Learning, a member of the National Academy of Education, and a Fellow of the American Association for Artificial Intelligence. He serves on numerous advisory boards and boards of directors. He has also published nearly 100 papers in scientific journals and the books *Seeing Differently: Insights on Innovation* and *The Social Life of Information* (with Paul Duguid) (Harvard Business School Press). He was awarded the 1998 Industrial Research Institute Medal for outstanding accomplishments in technological innovation and the 1999 Holland Award in recognition of the best paper in *Research Technology Management* in 1998. Dr. Brown has a B.S. in Mathematics and Physics from Brown University, and an M.S. in Mathematics and a Ph.D. in Computer and Communication Sciences from the University of Michigan.

Marye Anne Fox is Chancellor of North Carolina State University. Previously, she served in numerous capacities at the University of Texas, including Vice President for Research and Director of the Center for Fast Kinetics Research. Dr. Fox has held numerous visiting appointments and has had extensive consulting experience throughout her career. She has also been a board member of many organizations, including the National Science Board; she was chair of the Federal Science and Technology guidance group (1998/1999); and is currently a member of COSEPUP and Co-chair of the Council of the Government-University-Industry Research Roundtable. Dr. Fox is known for her contributions to organic photochemistry and photoelectrochemistry. Her research interests include physical organic chemistry, organic photochemistry, organic electrochemistry, chemical reactivity in non-homogeneous systems, heterogeneous photocatalysis, and electron transfer in anisotropic macromolecular arrays. Dr. Fox earned her Ph.D. from Dartmouth in 1974. She is a member of the National Academy of Sciences.

Ralph E. Gomory has been President of the Alfred P. Sloan Foundation since 1989. He was Higgins Lecturer and Assistant Professor at Princeton University from 1957 to 1959. Dr. Gomory joined the Research Division of IBM in 1959, became an IBM Fellow in 1964, and Director of the Mathematical Sciences Department in 1965. He was made IBM Director of Research in 1970, and held that position until 1986, becoming IBM Vice President in 1973 and Senior Vice President in 1985. In 1986, Dr. Gomory became IBM Senior Vice President for Science and Technology. Dr. Gomory served on the President's Council of Advisors on Science and Technology from 1990 to March 1993, and he has served in numerous capacities for many other academic, industrial, and governmental organizations. He is a member both of the National Academy of Sciences and the National Academy of Engineering. Dr. Gomory received his B.A. from Williams College in 1950, studied at Cambridge University, and received his Ph.D. in mathematics from Princeton University in 1954. He has also been awarded a number of honorary degrees and prizes, including the National Medal of Science. Dr. Gomory served in the U.S. Navy from 1954 to 1957.

Nils Hasselmo is currently the President of the Association of American Universities. Previously, he held numerous positions at the University of Minnesota, including President (1989-1997), Vice President for Administration and Planning (1980-1983), and Chairman of the Department of Scandinavian Languages and Literature and Director of the Center for Northwest European Language and Area Studies (1970-1973). Dr. Hasselmo has also served as Senior Vice President for Academic Affairs and Provost at the University of Arizona (1983-1988) and held visiting appointments at the University of Wisconsin (1964-1965), Harvard University (1967), and Umea University in Sweden

(1977). Dr. Hasselmo has received numerous fellowships and awards and is a member of a number of professional and educational associations, including the Board of the National Merit Scholarship Corporation, the Council of Big Ten, the National Association of State Universities and Land-Grant Colleges, the Universities Research Association, and the Kellogg Commission on the Future of State Universities and Land-Grant Colleges. Dr. Hasselmo received his baccalaureate from Augustana College and Ph.D. in Linguistics from Harvard University.

Paul M. Horn is currently Senior Vice President, Research, of the IBM Corporation, a position he has held since 1996. In his 20 years with IBM, Dr. Horn has been a champion for translating technology research into marketplace opportunities—first, as a solid state physicist, and then followed by several key management positions in science, semiconductors, and storage. Prior to his current appointment, Dr. Horn was Vice President and Lab Director of the Research Division's Almaden Research Center in San Jose, California. Dr. Horn graduated from Clarkson College of Technology and received his doctoral degree from the University of Rochester in 1973. Prior to joining IBM in 1979, Dr. Horn was a professor in the Physics Department and the James Franck Institute at the University of Chicago. Dr. Horn is a Fellow of the American Physical Society and an NSF Graduate Fellow, and he was an Alfred P. Sloan Research Fellow from 1974-1978. He is a former Associate Editor of Physical Review Letters and has published some 85 scientific and technical papers. In 1988 he received the Bertram Eugene Warren award from the American Crystallographic Association. Dr. Horn is a member of numerous professional committees, including the Council on Competitiveness, the Government-University-Industry Research Roundtable, the Clarkson University Board of Trustees, the UC Berkeley Industrial Advisory Board, and the Board of the New York Hall of Science.

Shirley Ann Jackson has served as the 18th President of Rensselaer Polytechnic Institute since July 1, 1999. Previously, she was Chairman of the U.S. Nuclear Regulatory Commission. During her tenure with the Commission she enhanced the regulatory effectiveness of the 3,000-employee, $472-million agency. Prior to joining the NRC, she was Professor of Physics at Rutgers University and held research positions at Bell Laboratories, the

Fermi National Accelerator Center, the Stanford Linear Accelerator Center, and the Aspen Center for Physics. She holds a B.S. in physics and a Ph.D. in theoretical elementary-particle physics from the Massachusetts Institute of Technology. She is a member of the National Academy of Engineering.

Frank H.T. Rhodes is Professor of Geological Sciences and President Emeritus at Cornell University. Before assuming the presidency at Cornell in 1977—a position he then held for 18 years—Dr. Rhodes was Vice President for Academic Affairs at the University of Michigan for three years. He joined the Michigan faculty as professor of geology in 1968 and, in 1971, was named Dean of the College of Literature, Science, and the Arts. He was professor and head of the geology department and Dean of the Faculty of Science at the University of Wales, and has served on the faculty at the University of Illinois and the University of Durham. Dr. Rhodes received a bachelor of science degree with first-class honors, as well as a doctor of philosophy degree, a doctor of science degree, and a doctor of laws degree from the University of Birmingham, England. He went to the University of Illinois in 1950 as a postdoctoral fellow and Fulbright scholar. Dr. Rhodes was appointed by President Reagan as a member of the National Science Board, of which he is a former chair, and by President George H.W. Bush as a member of the President's Educational Policy Advisory Committee. He has served as Chair of the American Council on Education, the American Association of Universities, and the Carnegie Foundation for the Advancement of Teaching. He has also served as a trustee of the Andrew W. Mellon Foundation. Dr. Rhodes has published widely in the fields of geology, paleontology, evolution, the history of science, and education. He is a principal of the Washington Advisory Group, a member of the board of directors of the General Electric Company, and a member of the Board of Overseers of Koç University, Turkey. He is currently president of the American Philosophical Society.

Marshall S. Smith is Program Director for Education at the Hewlett Foundation, and Professor of Education at Stanford University. He has been involved in helping to shape the nation's educational policies, especially as they relate to equal opportunity and high standards. He served as Undersecretary and Acting Deputy Secretary of the U.S. Department of Education from 1993 to 2000. In these capacities, he was the Chief Operating Officer of the Department and the Chief Policy Advisor to the Secretary. Originally trained in statistical techniques for research, Dr. Smith has extensive knowledge of policy issues from his years of previous governmental and academic experience. This experience has included research on such topics as computer analysis of social-science data, early-childhood education, critical thinking, and social inequality; teaching positions at Harvard, Wisconsin, and Stanford; and six years as Dean of the School of Education at Stanford. Dr. Smith's current research interests include national and state educational policy, educational quality, challenging educational standards, imaginative use of technology for learning, and policy and practices in education in emerging nations. He has been a member of several organizations, including the National Academy of Education and the National Council on Education Standards and Testing, and he served as the chair of several committees, including the National Academy of Sciences' Board of International Comparative Studies in Education and the U.S. Government Subcommittee on Educational Standards. Dr. Smith obtained his baccalaureate, M.A., and Ed.D. in Measurement and Statistics (1970) from Harvard.

Lee Sproull holds the Leonard N. Stern School Professorship of Business at the Stern School, New York University. She is currently Director of the Stern School Initiative in Digital Economy, a comprehensive project combining educational programs, research, and industry partnerships. Dr. Sproull is an internationally recognized sociologist whose research centers on the implications of computer-based communication technologies for managers, organizations, communities, and society. She has conducted research on technology-induced changes in interpersonal interaction, group dynamics and decision making, and organizational or community structure. Dr. Sproull has been a Visiting Scholar at Xerox PARC, Digital Cambridge Research Lab, and Lotus Development Corporation, and has published the results of her research in eight books and more than 60 articles. She has held previous appointments as Professor of Management at Boston University and Professor of Social and Decision Sciences at Carnegie Mellon University. She holds a B.A. from Wellesley College, and an M.A. and Ph.D. from Stanford University. Dr. Sproull is a member of the Computer Science and Telecommunications Board of the National Research

Council and the advisory board of MentorNet, and is a former Trustee of the Computer Museum.

Doug Van Houweling has been President and CEO of the University Corporation for Advanced Internet Development (UCAID) since October 1997. UCAID is a consortium of U.S. research universities, in collaboration with private- and public-sector partners, currently engaged in the Internet2 project to advance networking technology and applications for the research and education community. He is on leave from the University of Michigan. Dr. Van Houweling has been active in inter-university initiatives, serving on the board of EDUCOM—a consortium of 450 universities that developed computer networks and systems for sharing information and resources—and as a founder of EDUCOM's Networking and Telecommunications Task Force. He has also served as a board member of the Interuniversity Consortium for Educational Computing. Prior to going to Michigan, Dr. Van Houweling was Vice Provost for Computing and Planning at Carnegie Mellon and Assistant Professor of Government at Cornell. He received his undergraduate degree from Iowa State University and his Ph.D. in government from Indiana University.

Robert Weisbuch is President of the Woodrow Wilson National Fellowship Foundation. He joined the Foundation after 25 years at the University of Michigan, where he served as Chair of the Department of English, Associate Vice President for Research, and Associate Dean for Faculty Programs and Interim Dean at the Rackham School of Graduate Studies. He is a graduate of Wesleyan University and holds a Ph.D. in English from Yale University. He has received awards both for teaching and scholarship at Michigan, and is the author of books on Emily Dickinson and the stormy relations between British and American authors in the 19th century. While Dean of the School of Graduate Studies, he established a fund designed to improve the mentoring of graduate teaching assistants, created humanities and arts awards for faculty, and made diversity an integral criterion in evaluating program quality. He also headed up a two-year initiative to improve undergraduate education.

Wm. A. Wulf is currently on leave from the University of Virginia to serve as President of the National Academy of

Engineering. During 1988-1990, Dr. Wulf was Assistant Director of the National Science Foundation, where he headed the Directorate for Computer and Information Science and Engineering. Prior to joining the University of Virginia, Dr. Wulf founded Tartan Laboratories and was a professor at Carnegie Mellon University. While at Carnegie Mellon and Tartan, Dr. Wulf helped found the Pittsburgh High Technology Council and served as its Vice President and Director. His breadth and depth of experience have given him a unique perspective on the relationships between universities, industry, and government. Dr. Wulf is a member of the National Academy of Engineering, a Fellow of the American Academy of Arts and Sciences, and a Fellow of three professional societies (ACM, IEEE, AAAS). He is also the author of over 80 papers and technical reports, has authored three books, and holds two U.S. patents.

Joe B. Wyatt is Chancellor Emeritus of Vanderbilt University. Much of his earlier career focused on computer science and systems, in both industry and academia. In addition to holding faculty positions, he was also associated with EDUCOM in various capacities, including service as President and CEO. In 1976, he was appointed Vice President for Administration at Harvard and was named Chancellor of Vanderbilt in 1982, stepping down in 2000. He holds degrees in mathematics from Texas Christian University and the University of Texas. Mr. Wyatt has carried out research on behalf of the National Science Foundation, the Ford Foundation, the Office of Naval Research, and the Eli Lilly Foundation. He is coauthor of the book *Financial Planning Models* and the author of numerous papers and articles in fields relating to technology, management, and education. Additionally, Mr. Wyatt serves on a number of corporate boards, as well as professional and service organizations, and was a founding director of the Massachusetts Technology Development Corporation. He is Chairman of the Universities Research Association and past Co-chair of the Government-University-Industry Research Roundtable, as well as past Chairman of the Nashville Area Chamber of Commerce. Mr. Wyatt is also a member of the Association of American Universities, the Business Higher Education Forum, the Advisory Committee of the Public Agenda Foundation, and the Council on Competitiveness.